the **BIG** picture

reflections on science,

humanity, and a quickly

changing planet

the **BIG** PICTURE

David Suzuki
and Dave Robert Taylor

 David Suzuki Foundation

GREYSTONE BOOKS

D&M PUBLISHERS INC.
Vancouver/Toronto/Berkeley

Material in this book has been adapted from an essay series written by
David Suzuki and Dave Robert Taylor for the David Suzuki Foundation
and published in various publications between 1999 and 2008.

09 10 11 12 13 5 4 3 2 1

Greystone Books
A division of D&M Publishers Inc.
2323 Quebec Street, Suite 201
Vancouver BC Canada V5T 4S7
www.greystonebooks.com

David Suzuki Foundation
219–2211 West 4th Avenue
Vancouver BC Canada V6K 4S2

Library and Archives Canada Cataloguing in Publication
Suzuki, David, 1936–
The big picture: reflections on science, humanity, and a quickly changing planet /
David Suzuki, David Robert Taylor.
A collection of columns published under the title Science matters.

ISBN 978-1-55365-397-4

1. Human ecology. 2. Social ecology. I. Taylor, David Robert, 1971–
II. Title. III. Title: Science matters.

GF41.S89 2009 304.2   C2008-906316-3

Editing by Kathy Sinclair
Cover design by Jessica Sullivan and Naomi MacDougall
Cover photograph by Shannon Mendes Photography
Text design by Naomi MacDougall
Printed and bound in Canada by Friesens
Printed on acid-free paper that is forest friendly
(100% post-consumer recycled paper) and has been processed chlorine free.
Distributed in the U.S. by Publishers Group West

We gratefully acknowledge the financial support of the Canada Council for the Arts,
the British Columbia Arts Council, the Province of British Columbia through the Book
Publishing Tax Credit, and the Government of Canada through the Book Publishing
Industry Development Program (BPIDP) for our publishing activities.

CONTENTS

· · ·

PREFACE · 1

1 Blinded by Science · 5
   Research and the perils of ignorance

2 Smarter Than Your Average Planet · 34
   Interconnections in the biosphere

3 Getting to Know the Joneses · 58
   Protecting the diversity of life on Earth

4 Putting Mother Nature on the Payroll · 84
   Natural services and economics

5 Hot Hot Heat · 110
   Global warming and climate change

6 You Can't Get There from Here · 136
   Car culture and global transportation

7 Jellyfish—It's What's for Dinner · 163
   Feeding the planet in the twenty-first century

8   The True Cost of Gadgets · 194
    Technology and the culture of consumerism

9   Lights, Camera, Sound Bite · 219
    Social change and the media

10 Mr. Smith Goes to Washington · 245
    Public policy for a sustainable planet

    FINAL WORDS · 271

    ACKNOWLEDGMENTS · 273

    INDEX · 275

PREFACE

What a fascinating time this is.

A S A SPECIES, WE have come so far so fast that it simply boggles the mind. Not long ago, few people would ever travel more than a hundred miles from their place of birth. Today, even the most exotic location is just a plane flight away. A few centuries ago, we would meet maybe one hundred people over our entire lives. Today we can "meet" that many on Facebook in an hour—and then keep in touch with them—virtually anywhere in the world through smartphones and the internet.

Not all of us have such luxuries, but those of us living in developed countries should consider ourselves very lucky indeed. Stable governments, a reliable climate, and a relatively healthy planet have enabled our societies to take full advantage of all that the natural world provides. Using science and applying it through technology, we have made tremendous advances in medicine, physics, computers and telecommunications, biology, genetics, and much more. Today we are able to lead healthier, longer lives than human beings ever have in our history as a species. With the aid of marvelous technologies, we can now explore the depths of our oceans, manipulate life, and peer deep into the cosmos, pondering some of our most profound questions. Our ancestors would simply marvel at the world we've created. It really is an incredible time to be alive.

But with all this excitement, with all this hoopla and giddy enthusiasm, there is a trade-off. Modern life can be overwhelming. We are awash in unfiltered and often contradictory information on everything from the state of our planet to the state of our economy. Housing prices are down—that's bad! But housing is more affordable, so that's good! The stock market has collapsed—that's bad! But that's reduced the demand for oil, which means we're burning less of it and releasing less greenhouse gas pollution. That's good! Unless lower oil prices actually discourage energy efficiency and slow the transition to clean energy sources. That's bad!

It can be confusing, to say the least—if not downright paralyzing. Putting all these pieces together is a daunting task and one not easily achieved. In Vancouver, Canada, there's a television commercial for a local news channel that uses a version of the slogan "Helping you make sense of your world." That would be wonderful if it were true. Sadly, it's mostly ironic, for nowhere is our world more fractured and chaotic than on the television news. If you wanted to make sense of things, the last place you would likely find enlightenment is in a television newscast. The nature of the medium provides little room for context.

But the media aren't our only source of information overload and it's just one part of the story. The fact is our modern world is very complicated. Everything now takes place at a global scale—politics, trade, war, economics, technology—and all seem to be moving faster and faster with every passing year. Yet all of these things that so dominate our headlines and our lives—all of these things are merely social constructs created by humans. Ultimately, they still depend on something else: the continued existence of a healthy and stable planet to provide us with the basics of life. Without a healthy planet to cleanse our wastes and provide us with resources, we will wither. Unless we can find a way to live in balance with the natural systems that sustain us, our species will ultimately reach a dead end.

That's the bottom line.

So what do we do? We live in a fractured world, pummeled by random information, very little of which actually addresses this bottom-line reality. We hear doom and gloom about the environment virtually every day, but then we hear that fixing the problems is too expensive. We read about the need to "jump-start" the economy, but nobody asks "to what end?" or questions how it's actually possible to achieve unrelenting economic growth on a finite planet. It's enough to make you shout, "Enough information already! Where do we go for context?"

This book, a selection of essays compiled to help readers put some of the fractured pieces back together, represents a modest start in that direction. It's not comprehensive, by any means. In truth, it will probably raise more questions than it answers. But perhaps that's a good thing, for such questions are not asked often enough in our universities, on the evening news, or by our elected leaders. That in itself says something about the pace of our modern information age and the deep and resounding need for context—the need to step back from the chaos, take a deep breath, and try our very best to see the big picture.

1

BLINDED BY SCIENCE

Research and the perils of ignorance

SIR JOHN MADDOX, A noted chemist and former editor of the prestigious journal *Nature*, once stated: "The questions we do not yet have the wit to ask will be a growing preoccupation of science in the next fifty years."

A thought-provoking statement, to be sure. And one that neatly sums up all that is both wondrous and dangerous about science. Wondrous, because science is a journey without end—and an unpredictable one at that. Dangerous, because this means we are always operating at a certain level of ignorance. Science is not a game of absolutes. And humans tend not to deal well with shades of gray.

On a personal level, science means the world to me. It is, after all, how I discovered nature. As a boy, I simply explored the world and wondered at the diversity of it all—the beaches, insects, mountains, forests, fish, and ponds. All evoked wonder and curiosity, inspiring me to learn more and providing the first hint of a career in science. As a man, I became fascinated with genetics and was moved to dig deeper, exploring the roots of our identity, our origins, and our migrations. On a much larger scale, science has fundamentally changed the planet and how we see it. Indeed, science, and the application of it through technology, have created a kind of world about which I could never have dreamed when I was growing up.

As powerful and as useful as it is, science is one-dimensional. It is elegant but imperfect. It offers us a way of thinking and a logical method of observation and repetition that give us insight into the world around us. But because of its reductionist nature, science can never provide us with a complete understanding of how the world works. One of the hallmarks of science is that experiments must be repeatable. So when performing experiments, we remove all confounding factors that could influence or confuse the results. But nature doesn't work that way. Nature does not operate in a vacuum. Interconnections among the various parts of the natural world are what actually drive it. When we pull it apart, we lose context—and that can mean everything.

Recognizing the limitations of science, however, does not negate its value. Nor should it push one to extremes. It doesn't mean that we are somehow ignorant of the world or that science can't be trusted. The great strength of science is that it gives us the capacity to probe nature and learn its secrets. Through science we have learned to split atoms and release energy, read DNA, and synthesize genes. And we learn more every day. But we also should not blindly accept every new discovery as gospel or every new technology as a savior. This is especially important to remember in new, revolutionary areas where experiments and observations constantly cause reevaluation, even rejection, of hot ideas.

Our species is really quite special. We've learned so much and come so far in such a short time. So far, in fact, that it's all too easy to fall into the trap of hubris—of thinking that we really understand this world and that we can fix any problem that might come up. That's a dangerous assumption.

I have a great deal of faith in humanity's capacity to solve problems. I've witnessed it firsthand throughout my life, from antibiotics that brought me back from near death when I was a boy with pneumonia to the space race that led to astounding leaps forward in computer technology, telecommunications, and genetic engineering. But as a society and as a species, we've become so used to science and technology that we've forgotten that these are just ways of understanding and manipulating the world. They do not solve problems on their own. And as many problems as they do help resolve, they also create them anew.

As you will read for yourself in these essays, science changes. Assumptions are challenged, hypotheses disproved. But the beauty of science is that when we're presented with new information, it allows us to change our minds. It gives us a rational explanation as to why the world doesn't work the way we previously thought it did. Certainly, these new answers always bring up more questions. But no one ever said that life was easy—or easy to understand.

Science proceeds not in a beeline, but at a stagger, stumbling home in the darkness after a night at the pub, drunk on its own discoveries. Tomorrow, it may find that many of its hypotheses were dead wrong. But tonight it will celebrate all it has learned and raise a toast to the questions it does not yet have the wit to ask.

THE BEAUTY AND THE HORROR OF SCIENCE

At an international biotechnology conference in Vancouver, Canada, an industry spokesperson made reference to the hundreds of protestors outside and suggested that biotechnologists had obviously done a poor job convincing the public about the benefits and safety of their products. Thus, she trivialized the opponents' concerns as based on ignorance and not deserving serious attention.

It's unfortunate that GMO (genetically modified organism) has been used to refer to foods created by inserting genes from one species into another. I say "unfortunate" because for the past ten millennia human beings have been genetically modifying plants and animals by selection and breeding. All of the food we eat was once wild and, whether it's corn, rice, or chickens, we have dramatically increased yields and changed characteristics by genetic modification. Even more remarkable, the array of dog breeds—from Chihuahuas to Great Danes— were all derived by breeding from tamed wolves.

Critics of biotech food have labeled them "Frankenfoods"— an allusion to the famous story by Mary Shelley. We often forget that Frankenstein was the doctor/scientist, not the monster he created. The story is an apt allegory for the powers we have come to apply with biotechnology.

Victor Frankenstein was involved in experiments to find the secret of life. We watch in horror as he is driven by his curiosity to solve the mystery. Yet, one of the enchanting attributes of scientists is that capacity for enthusiasm and single-minded focus. As Theodore Roszak has written:

> It is both a beautiful and a terrible aspect of our humanity, this capacity to be carried away by an idea. For all the best reasons, Victor Frankenstein wished to create a new and better human type. What he knew was the secret of the

creature's physical assemblage; he knew how to manipulate the material parts of nature to achieve an astonishing
result. What he did not know was the secret of personality
in nature. Yet he raced ahead, eager to play God, without
knowing God's most divine mystery.

One of the most horrifying things I have ever witnessed
was an experiment in which a cat was *decerebrated,* that is,
it had all of its brain scraped out. It was still alive, and when
an electrode was inserted into a certain part of its brain stem,
the cat began to walk on a treadmill. It was a macabre experiment, but the scientist's enthusiasm in concluding that the
nerves controlling a cat's walking ability must reside in the
spine blinded him to the horror of what he was doing.

I spent twenty-five years running a genetics lab, studying
how genes control an organism's development and behavior.
The great joy of the lab for me was the excitement and exhilaration of research and the moments of sheer ecstasy when
we gained a new discovery or insight. Yet when we wrote
experiments up for publication, all of that joy and emotion
were expunged. We would never get a work published if we
included a description of the "breathtaking beauty of the vivid
scarlet sheen of exquisitely arranged rows of ommatidia" of a
fly's eye or the exhilaration upon recovering a mutation induc-
ing paralysis at different temperatures. Yet that's why we were
hooked on the work.

The reason we can't express emotion is because science's
great boast is objectivity. Ever since Descartes and Newton,
we have tried to separate ourselves from the object of study,
tried to focus on a part of nature, measuring or describing it
in mathematical or chemical terms. In the process, we have
acquired profound understanding of some of the most basic
parts of the cosmos—subatomic particles, atoms, genes, and

cells. But by focusing on the parts, we frequently lose sight of the whole—of patterns and rhythms that make the quest interesting in the first place. And that's often what the public senses is wrong with scientists.

So even though she may have had the best of intentions, when that biotechnologist in Vancouver trivialized legitimate concerns as being merely ignorant, she revealed the very attributes that the public fears about science—the single-mindedness that can turn a scientist into a Frankenstein.

CAGED ANIMALS CAN GO STIR-CRAZY

One of the hallmarks of the scientific method is that experiments must be standardized and repeatable. But what if, in the push to "standardize" and remove all potentially confounding factors, we've created conditions so unnatural that our experiments themselves become invalid?

It's a question that's long overdue. After all, we once thought that we could study an animal like a chimpanzee in a zoo or cage to learn all about it. But when Jane Goodall studied them in the wild, she found they were very different. A few years ago, researchers at the 35th Congress of the International Society for Applied Ethology (the study of animal behavior) began questioning the validity of "controlled conditions."

Scientists routinely use laboratory animals, particularly mice, to test new medicines, medical procedures, drugs, foods, and more. To reduce the number of variables between animals and, scientists hope, to obtain more accurate study results, the animals are genetically homogenous and raised in standard clinical cages, with few stimuli other than the experiment itself. This clean, sterile environment is considered healthy and normal.

But in 1996, a Swiss animal behaviorist noticed that mice and rats reared this way might actually be decidedly abnormal. Using infrared cameras to spy on the nocturnal lab mice in the dark, researchers found that most of them behaved very strangely after their handlers had gone home for the day. The mice continuously repeated seemingly meaningless behaviors, such as cage biting and cage scratching. Such repetitive actions by animals are called "stereotypies" and they are often considered signs of boredom or stress. Still, the mice were thought to be basically "normal," and their actions attributed to mere habit.

In humans, however, the presence of stereotypies is thought to indicate damage to part of the brain called the basal ganglia, which are believed to regulate how we initiate movement. Evidence suggests that stereotypies may actually indicate permanent brain damage in other animals as well. In September 2001, the journal *Nature* reported a University of California study that examined parrot behavior using a procedure that is normally used to test for damage to human brains—particularly to the basal ganglia. Parrots that exhibited stereotypies failed the test, indicating potential damage to this region.

What about science's favorite test mammal—the lab rat? Rodents spend a great deal of their lives searching for food and building nests. Denied an outlet for such basic instincts, the animals suffer stress and possibly impaired brain function. Indeed, studies have found that rodents kept in enriched environments perform better in memory tests.

If test subjects are already impaired, scientists may be inadvertently invalidating studies by relying on data from brain-damaged animals. As behavioral scientist Joseph Garner points out, "if the abnormal behaviors that are so common in captive animals are indicative of abnormal brain function, then these animals can hardly be considered to be good models of normal animal or human functioning."

This also raises questions about the validity of studying the behavior of large mammals such as killer whales in captivity, where they are confined to small pools and receive little natural stimulus. In the wild, killer whales can travel great distances every day, moving and hunting in family groups. It should be obvious that the behavior of such animals in captivity will not look anything like the true behavior of their wild counterparts.

Clearly this issue needs to be examined further. Ethically, we must ensure that test animals are treated as humanely as possible. Practically, if the animals that we use for experiments are already impaired, then are the conclusions we draw valid?

There may be simple ways to solve the problem. A preliminary study done at Utrecht University in the Netherlands examined ways to create a more natural environment for test rodents, such as adding nesting material and scattering feed so rodents can forage. It found that such steps could be taken without jeopardizing experiments. In a more comprehensive 2004 study, researchers from the University of Zurich tested 432 female lab mice in both enriched and standard environments. Their findings, published in *Nature,* concluded: "The housing conditions of laboratory mice can be markedly improved without affecting the standardization of results."

So what are we waiting for?

.   .   .

WE MUSTN'T NEGLECT THE BASICS OF BIOLOGY
There's an old stereotype of biologists as bespectacled eccentrics working in the field to diligently catalog all sorts of creatures, from worms to butterflies and birds. Unfortunately,

the reality is that such descriptive scientists have all but disappeared, and that's not good for the environment or us.

Scientists tell us that the Earth is currently undergoing the sixth great mass extinction in the planet's history—the biggest since the dinosaurs died out sixty-five million years ago. But this event is also the first of its kind—a mass extinction caused not by natural climate shifts or an enormous meteor from space, but by one of the planet's own species—humanity. This has scientists worried because a diversity of life forms is extremely important for ecosystems to provide us with services, like cleansing our air and water, creating new soil, and capturing the sun's energy through photosynthesis.

The problem is we can't save or even appreciate species that we don't know exist. And while scientists were once driven by the desire to understand the basics of the world around us, like what sorts of species we share the planet with, they are now influenced more by fads and sexier areas where there is more funding—like cancer research and biotech. Of course, such research is important, but right now the basics are being ignored. Descriptive sciences like taxonomy (the discovery and classification of living things) have been very poorly supported for years. Dr. Geoffrey Scudder, an entomologist and former head of zoology at the University of British Columbia, once told me that half the insects in the Canadian national collection in Ottawa have never even been classified! Taxonomy even gets little respect from within the scientific community. An article in the journal *Science,* for example, pointed out that the top five taxonomists in the world rarely have their papers published in influential science journals, and instead seem relegated to relative obscurity. Essentially, we are operating in ignorance. Microorganisms, for example, may be more numerous in species and certainly greater in total biomass (the weight of living matter) than all the other species combined.

And we know their work is a critically important part of the engine of nature. Yet we know next to nothing about them. In order to identify them, we need morphological or molecular markers (ways to tell them apart), but to obtain those, we have to be able to grow the microorganisms in a lab. Unfortunately, we have little idea of the needs of all microorganisms in terms of a culture medium or conditions—in other words, what it takes for them to survive and grow in a lab setting. So what we manage to isolate from a pinch of soil may be a tiny fraction of all that exist.

The same is true of microorganisms in our oceans. Until 1988, biologists believed that plankton were the base of the marine food chain. Then picoplankton, which are ten times smaller than plankton and pass right through plankton nets, were discovered. Now it's thought that picoplankton are far more numerous and important than plankton in terms of overall ecosystem function.

Estimates of the number of species on Earth run between 2.5 million and 200 million, with 10 to 20 million being the most common assessments. Yet all we have actually identified to date are 1.5 to 1.8 million—or about 10 to 20 percent of all living things. How can we presume to "manage" the planet's natural resources when we have such a poor inventory of its constituents and a virtually useless blueprint of how all the components interact?

Yes, there are good reasons why sexy fields of science are popular. But for these fields to succeed and advance, they need to be built on a solid foundation of basic science. Without a firm grasp of the basics of life on Earth, we could lose most of it. And we'll be left groping blindly for answers when we don't even know if we're asking the right questions.

FUSION CLAIM SPARKS FEAR, HOPE

When researchers in Tennessee zapped a chamber of organic solvent with high-speed neutrons, they hoped it would trigger a fusion reaction. Whether they succeeded is debatable, but they certainly ignited a furor in the scientific community, much of which was misplaced.

Nuclear fusion, whereby atomic nuclei slam together at high temperatures, fuse, and release a great deal of energy, is sort of the Holy Grail of energy production. It's this reaction that powers the sun and gives the explosive force to a hydrogen bomb. If we could harness this power, it would solve many of the energy problems that plague us today. (Although it would create new problems—imagine how much faster we could consume natural resources if we had an unlimited energy supply!) But a controlled fusion reaction potentially suitable for power generation has thus far eluded scientists.

That is until 2002, when researchers at the Oak Ridge National Laboratory claimed that they had produced "bubble fusion" in a beaker. The process centers on the rapid collapse of bubbles formed by sound waves in a liquid. Under the right conditions it is possible for these bubbles to implode with enough energy to induce fusion. The researchers at Oak Ridge claimed to have seen evidence of such a process.

Others, however, believed the experiments were flawed and thus invalid. When the respected journal *Science* decided to publish the findings, it created a great deal of consternation in the scientific community. Many researchers were worried that another debacle like "cold fusion" would result. You may remember that one from 1989, when researchers insisted they could get a fusion reaction from a blob of palladium metal in a test tube. Those claims later proved to be unsubstantiated.

Naturally, physicists were skeptical about claims involving "tabletop" fusion. No one wanted to see hopes raised, only to have them quashed if the experiments could not be repeated or if the results could be explained by some miscalculation. And some expressed concerns about the reputation of the discipline. At a time when funding dollars were spread thin, an embarrassment could prove costly.

Such concerns and fears are understandable. As the former head of a large genetics lab, I know the practical importance of funding, publishing, and getting results. But we have to be careful not to cross the line from scientific diligence to censorship. We must never lose sight of why we are conducting experiments in the first place—hope. Science is hope—the hope of new discoveries, the hope of making a better tomorrow, of improved health and long life, of understanding our past, our present, and our future.

Unfortunately, science today is also very political. It's too important and too expensive not to be. Everyone has a finger in the pot and wants his or her fair share of the glory. And everyone wants to distance himself far away from failure. That's what most of the fear about the fusion paper was all about. Physicists peppered *Science* with letters and emails asking the editors to pull the paper from publication. Even the Oak Ridge lab administration withdrew its support and tried to get *Science* to rethink its decision. But the journal held firm.

Bubble fusion has never been replicated. And, in 2008, the lead scientist involved in the fiasco was reprimanded for misconduct. Yet none of this proves that the journal made the wrong decision in publishing the original paper. Assuming the paper met its standards for publication, had the journal bowed to pressure, it would have been a sad day for science. Indeed, it would be tragic if scientific journals succumbed to peer pressure and did not publish any papers that might be wrong or make them look bad. Science is all about making mistakes. That's how we learn. If scientists are so concerned about their reputations that they don't report exciting findings because their ideas may not be in vogue or for fears that they could be incorrect, then the scientific endeavor becomes fettered and we all lose out.

Science is a process of discovery. That's how it moves forward. We build on both the mistakes and the successes of those before us. Today, when I tell students what the hottest genetic theories were thirty years ago, they crack up. Sometimes, we can only get things right by getting them wrong first.

*   *   *

SPACE IMAGES PEER INTO THE PAST AND THE FUTURE

Scientists were thrilled when newspapers around the world first published spectacular images taken with the Hubble Space Telescope's new advanced camera. The pictures showed us never-before-seen glorious spiral galaxies and ethereal nebulae—stunning views of places far, far away. But for many, the big question is this: is there anyone else out there looking back at us?

I hope so. The Hubble pictures, while beautiful, are also lonely. If you're like me, you probably overlooked their significance, paused only briefly to scan them, then quickly moved on to deal with something else in life. Since 2002, Hubble has amassed thousands more astonishing pictures of our universe. But we're so used to seeing similar computer-generated images on television and in science fiction movies that we hardly think twice about such things anymore.

Seeing pictures like these, however, is truly a privilege. Think about this: we are the first generation of humans ever to peer deep into space. For almost all of human existence, we've been limited to staring at the night sky with naked eyes. It's been less than four hundred years since Galileo first peered through his blurry prototype telescope, and not quite two decades since the launch of Hubble—the first space-based telescope.

Free from atmospheric distortion, 350 miles above Earth, Hubble has allowed us to probe the depths of the universe. These views are not just of distant places, but of distant times, since the light reaching the telescope may have taken millions or even billions of years to get here. In some cases, we are witnessing events that occurred before Earth even existed!

As spectacular as the images are, what everyone really wants to see is something like another Earth—another planet as lush and watery and green as ours that would be capable of sustaining complex life forms. Something to show we aren't alone. But even Hubble can't do that.

Right now we have to rely on an educated guess. Our galaxy hosts about three hundred billion stars, many of which have solar systems, some of which may have planets suitable for life. According to a famous formula developed by astronomer Frank Drake, we can guess the number of planets that could harbor intelligent life by multiplying eight variables. Depending on the assumptions used, the Drake equation can come to either wildly optimistic or pessimistic results.

Some scientists favor the pessimistic view, arguing that a planet like ours is probably exceptionally rare in the universe. Moreover, just because a planet is capable of sustaining life does not mean that life would necessarily appear or that it would be provided with just the right conditions necessary to evolve into something intelligent.

Others, however, say that if a planet has the properties necessary for life, then life would be virtually inevitable. In a 2002 article published in *Nature,* two Australian physicists argued that if suitable planets exist, the odds of life appearing on them are high because life here originated so rapidly. The time from when conditions were first amenable to life to the time life emerged on Earth was less than five hundred million years—speedy in geological terms. Factoring that rapidity into the Drake equation, the researchers say that if there are other planets like Earth out there, odds are that at least one in three will harbor life.

So far, astronomers have found more than seventy planets in solar systems other than our own. But they are all lifeless gas giants, much like our Jupiter—some even larger. Finding smaller, Earth-like planets closer to a star is a much more difficult task. Until 2007, NASA had plans to launch a Terrestrial Planet Finder, which would have been able to pinpoint Earth-like planets, in 2013. Sadly, those plans are now on hold due to budget cuts. However, NASA still plans to launch a next-generation space telescope, the Kepler Mission, sometime in 2009. Kepler was to be the precursor to the Terrestrial Planet Finder by narrowing down areas for study that have good odds of harboring Earth-like planets. Although not as detailed as the Terrestrial Planet Finder, it will still be the first mission capable of finding Earth-sized and smaller planets. Such technologies certainly won't tell us all we want to know about our universe, but they will take us another step towards answering some of our oldest questions.

GENOME SEQUENCE JUST THE BEGINNING

As a geneticist, I could not help but feel a sense of wonder and awe after the 2001 release of the completed human genome sequence. For most of my forty years as a geneticist, I never anticipated that this would occur in my lifetime. The determination of the sequence of all three billion letters in the DNA blueprint of human cells was a stunning technical achievement. Now the really interesting and difficult work of trying to make sense of it all is underway. That task will take much, much longer.

The human genome is incredibly complex, and the genome project has elucidated just 2 percent of it (the other 98 percent is highly repetitive sequences of what is currently thought to be non-coding "junk" DNA, the function of which has yet to be determined), but that didn't stop some newspapers from running misleading headlines such as "Revealed: The secrets of who we are." Others turned this scientific advance into a triumphant business story with headlines like "Private sector wins genetic-code race."

For the most part, however, newspapers got the story right—pointing out that the sequenced human genome has opened up more questions than it has answered. For example, although it now appears that humans have fewer than twenty-five thousand genes (the lowly fruit fly has more than thirteen thousand), we have little idea how these relatively few genes are harnessed and orchestrated to transform a fertilized egg into an organism as complex as a human being. It suggests that the basic biological differences between fruit flies and human beings are not nearly as great as differences in mere size or appearance would suggest.

In fact, comparisons between the human genome and the sequenced genomes of other species reveal remarkable genetic similarities. For example, we share as much as 10 percent of our genes with much simpler organisms like the fruit fly and the roundworm. We also share some 233 genes with a bacterium—and those genes are not found in the genomes of the fruit fly, the roundworm, or yeast. So at some point in our evolutionary history, our ancestors must have taken on bacterial genes, or vice versa.

Even more enlightening is the comparison of our genes to other, more closely related species, such as mice, chimpanzees, and even Neanderthals. Determining the mouse genome was important because of its similarities to the human genome. Genetic experimentation with mice could yield important medical advances for humans in the future. Humans and chimpanzees share about 99 percent of the same DNA, so a comparison of the two genomes should inform us about what genetic factors make us human. And examinations of DNA from other primates can help us uncover where and how they split from us on the evolutionary tree.

Still, in comparing the genomes that have been sequenced so far, it is the similarities that are most striking. Evolution has been remarkably efficient, exploiting the same genes over and over throughout life. The fact that a few genetic differences are all that separate humanity from the rest of the animal kingdom ought to give us all a much-needed dose of humility. Similarly, the finding that genetic differences within racial groups are greater than those between racial groups strikes a final nail in racism's coffin. Genetically, race consists of superficial physical similarities and little more. The Human Genome Project confirms that all life on Earth is genetically related through our shared evolutionary ancestry. Other species are our kin, presenting a very different relationship with us than if they were just resources at our disposal.

With the human genome sequenced, scientists have begun the vastly more complicated process of determining what the three billion bases that make up our genes do, how and when they are turned on and off, and how the proteins they design collaborate to carry out various functions. That is the real challenge, for the sequenced human genome merely represents a list of parts. As geneticist Eric Lander has said: "We've called the human genome the blueprint, the Holy Grail, all sorts of things. It's a parts list. If I gave you the parts list for a Boeing 777 and it has one hundred thousand parts, I don't think you could screw it together, and you certainly wouldn't understand why it flew."

.   .   .

HUMAN GENOME CONTINUES TO SURPRISE

Imagine discovering that the person running your favorite Fortune 500 company was not the CEO, as everyone presumed, but rather the bicycle-courier guy in spandex shorts and a goatee who everyone thought just delivered the messages.

That's pretty much how scientists working on the ENCODE project must have felt after analyzing the first part of the human genome.

ENCODE, short for Encyclopedia of DNA Elements, is a massive project that aims to catalog all of the functional elements of the human genome. The recently completed first stage of ENCODE cataloged just 1 percent of our genetic code, but that represents some thirty million bases or "letters" of DNA, in this case chosen randomly from forty-four different parts of the genome. Analyzing that 1 percent of our genetic structure took 308 scientists from ten countries four years to complete.

All that effort has uncovered something marvelous: what I and other geneticists took for granted for decades may have been wrong, or at least a wild simplification of what's actually going on.

Until very recently, accepted dogma in genetics was that DNA, specifically DNA in the form of genes, contained all the instructions necessary to make proteins. These proteins then made things happen at a cellular level; thus a gene is "expressed" and its instructions carried out. Another chemical, called RNA, was like a Xerox copy that simply replicated information from the DNA and transferred it to the area where proteins are made, shuttling information back and forth like a courier. It's a nice, tidy explanation for a complicated process. And in hindsight, it's probably a little too tidy.

Scientists first came up against the limits of this explanation when they mapped the human genome. To their surprise they found that people have only some twenty-two thousand protein-encoding genes. Yet organisms like *C. elegans,* a tiny worm, or my specialty, the fruit fly, have almost as many— nearly twenty thousand of them. If these genes are providing all the instructions on how to build and maintain an organism, how can such obviously more complicated creatures like humans have similar numbers of genes to simpler creatures like insects?

One answer may be found in the majority of our DNA that does not, as far as we know, code for proteins—what scientists used to call "junk." When ENCODE researchers started their project, they probably assumed that because only a small fraction of our DNA coded for proteins, only a small fraction of whatever they looked at would be transcribed into RNA, the messenger that delivers the instructions on how to make the protein.

Instead, ENCODE researchers found that much of the human genome is transcribed into RNA. It's just that the information contained in it isn't necessarily read to make proteins. So then what is the role of junk DNA, and what does all this extra RNA do? As yet, no one really knows, but it's clear that the human genome is much more than the sum of its genes. In fact, genes themselves may actually take a backseat to RNA in the development and functioning of an organism.

It's amazing for me to look at what we know now compared to when I ran a genetics lab back in the 1960s and '70s. In fact, when I tell students what we used to think back then, they can't help but giggle at our naïveté. I may be overstating the role of the bicycle courier in my Fortune 500 company analogy. But I could be understating it too. It's still too early to say whether RNA—our genetic bicycle courier—is actually running the show or not. But what has become clear is that there are a lot more bicycle couriers running around out there, delivering much more information than seems necessary and perhaps even making decisions on the fly. They may not necessarily be running the company, but they certainly have the ear of whoever does, and they aren't keeping their opinions to themselves.

.　　.　　.

SCIENCE NEEDS A MAKEOVER

Science has an image problem. Not that science isn't important to people today. Indeed, science and technology influence our lives now more than ever before. The problem is in the way scientific issues are often portrayed and communicated— both to the public and to other scientists.

A common complaint from scientists is that the media get everything wrong. They sensationalize. They oversimplify. They draw unwarranted conclusions. It's not surprising that with such an attitude, scientists are often skeptical and reluctant to talk to the media.

Many scientists want more control over their stories. A European survey found that 90 percent of scientists polled believe reporters ought to provide full scientific details in their stories and allow scientists to make changes before they are published. Of course, most journalists would never accept such demands; nor should they. Scientists should not receive special treatment. Imagine the kind of reporting we would have if journalists allowed politicians to edit their stories!

Reporters, however, complain that science stories are often dull, irrelevant, or impossible to comprehend. In an analysis published in *Nature,* sociologist Donald Hayes used a formula called the LEX scale to rate readability of journals. On this scale, the lower the number, the easier something is to read and the more understandable it is. Dr. Hayes found that children's books have a LEX score of about −32 (easy to read). Newspapers, on average, have a LEX score of zero. Back in the 1940s, science journals also scored about zero. Today, these journals reach LEX scores well into the +30s and beyond, meaning they are very difficult to understand.

In fact, journals today are often so loaded with jargon that scientists themselves have trouble reading them. This means there's a far greater chance that reporters and the public will misunderstand a story or never get the information at all. And this knowledge gap is getting worse, despite the ubiquitous impact of science and technology in our daily lives.

Look at the portrayal of many important environmental issues in the news. Anyone who regularly reads science journals knows that the vast majority of evidence indicates that human activities are steadily eroding many of the planet's life-support systems. We are changing the climate, causing species extinction, and spreading pollution across the globe. Yet the mainstream media largely focus on the fantastic changes, the doomsday scenarios. Gradual erosion fades into the background, and as a result, public concern becomes aroused only when problems reach critical levels—when it may already be too late to do anything.

No single group is at fault for this growing problem. Most scientists simply don't receive communication training. Plus, newspaper chains are firing reporters en masse, compressing beats, and not giving reporters the time and resources they need to adequately cover science issues. Few reporters exclusively cover either science or the environment. As a result, only the sensational makes headlines and the public continues to perceive scientists as faceless folks in lab coats and to see science's relationship to humanity as chaotic and disconnected.

I think there is a true thirst for good, clear science stories. As children, most of us had an innate sense of curiosity that made us question everything about our world. Why is the sky blue? Why do birds do that? Why can't people fly? But many of us lose that curiosity as adults—as though the world is just too complicated to be bothered asking about it. Yet explain a science story in a simple way to adults, and they, too, will often become fascinated and share in the wonder of the world around us.

In 2001 a magazine called *Seed* started up, and it's still going strong. *Seed* attempts to popularize science and technology by focusing on easy-to-understand prose with a dash of sex appeal. Perhaps by making science more hip, we will be able to elevate the general level of scientific awareness in society. Such awareness will be crucial if we are to make the right decisions to lead us into a sustainable future.

SHADES OF GRAY DOMINATE SCIENCE

In a world where things are usually presented as either good or bad, us versus them, or black versus white, complicated issues about science and nature can leave the public confused, ambivalent, or both. And that's not good for anyone.

It's human nature to categorize. Our ability to group a complicated array of items and issues into conceptual categories helps us understand the world and enables us to strategize and perform complex tasks over time. At its most basic level, it has helped us survive.

One way to describe this ability is "framing." Frames are like mental shortcuts that take advantage of what we already know to categorize new information. The media take advantage of this inherent desire to categorize complicated issues quickly by presenting them in black-and-white terms. Politicians tend to do the same thing, presenting issues as though there are really only two choices—the right one and the wrong one. You're either with us or against us.

But nothing in nature is simple. Take the confusing issue of global dimming. Most people know about the problem of global warming—pumping out heat-trapping gases into the atmosphere has put the planet's greenhouse effect into overdrive. As a result, we've disrupted our climate—leading to rising temperatures, shifting weather patterns, and an increase in extreme weather events.

Global dimming, however, is another phenomenon, relating to the amount of sunlight penetrating the atmosphere. This amount has actually decreased by about 5 percent since the late 1950s because of all the light-blocking soot we've been putting in the air from fires, smokestacks, and tailpipes. But lately, studies have shown that the Earth is actually brightening again—most likely because the amount of pollution in the atmosphere has dropped since the collapse of the Soviet Union.

That's good news—at least at first glance. Unfortunately, scientists also tell us that without this pollution in the air, more light and heat will get to the surface of the Earth. And since we have more heat-trapping gases in the atmosphere, that heat will tend to accumulate—potentially making global warming worse. Then again, that extra heat and sunlight might also increase evaporation and cloud formation, which could increase global dimming once again. It's complicated. There's also the confounding factor of China and India as emerging energy superconsumers. If they continue to expand their economies using dirty energy sources like coal, the resulting particulate matter in the air could also increase global dimming.

Climate contrarians—people who insist that, despite all the evidence, global warming is an urban myth—like to use this sort of complexity as evidence that scientists don't really understand our climate. In reality, it just doesn't boil down into a neat sound bite. Our climate is a very complex system that is influenced by a number of "climate forcing" mechanisms, including greenhouse gases and aerosols like soot in the atmosphere.

There's no shortage of challenging science topics—from climate change to stem cell research, cloning, euthanasia, and more. Genetically modified crops are hardly discussed in North America, where they are widely grown, but they have been hotly debated in Europe because they are created using a radical new technology, and many people feel that they should undergo rigorous long-term testing before being released into the environment.

However, genetic modification could also be beneficial. Farm research in China on genetically modified rice with built-in pest resistance has found that this particular variety can slightly improve yields. More important, it can greatly decrease the amount of pesticides farmers use. This seems like a great win, but no one knows how long this pest resistance will last or if it will cause any other health or environmental problems. And how does this system compare to organic farming, which uses no chemical pesticides and also has strong yields? Again, there are no easy answers.

Our tendency to frame complex issues as either good or bad is convenient, but it's preventing real debate from taking place. The sooner we stop trying to oversimplify these issues and develop a common language to discuss them, the better off we will all be.

·   ·   ·

SPEAKING OUT IN THE NAME OF SCIENCE

When should scientists be advocates? It's a simple question that often raises a storm of controversy.

Some argue that scientists should not be advocates, period. According to this view, science is value-neutral—simply a quest for knowledge. Scientists should conduct research to reveal information about our world but leave it up to society to decide what to do with that information.

Of course, such a viewpoint ignores the fact that no activity is truly value-neutral. Even deciding what research to undertake requires a value judgment. So for most people, the question is really, at what point should scientists take a stand on an issue?

Correcting misleading information in the media would be a good start. Right now, well-heeled groups that have a lot to gain from maintaining the status quo are actively funding campaigns of misinformation to confuse the public about science issues. Some of these campaigns are organized through conservative think tanks based in the United States. Their presence is felt in other countries too.

One of their most successful strategies in recent years was to have spokespeople consistently complain about the "liberal bias" in the media. It was like a mantra repeated over and over. Of course, there was no liberal bias, but by repeating the phrase ad nauseam, people began to believe it. They assumed it must be true. In response, media (in the U.S. in particular) took a sharp turn to the right.

Journalist Chris Mooney's book *The Republican War on Science* chronicles just how successful and far-reaching these groups have become. He argues that there has been a deliberate misrepresentation of science and an exaggeration of uncertainties that stretches all the way to the White House. From acid rain to climate change, and from birth control to endangered species, stem cell cloning, and more, Mr. Mooney says industry groups and the Bush administration deliberately tried to keep the public misinformed.

Don't think this doesn't happen elsewhere. Canada's federal government followed the Bush administration's lead, eliminating the important position of national science advisor, muzzling scientists working for the Ministry of Environment, and requiring science reports to be vetted through political staff before release. *Nature* published an editorial castigating this political interference with science.

Newsrooms across the country are routinely bombarded with articles from rogue scientists or "environmental consultants" who have a story to tell. These stories are usually the opposite of the prevailing scientific opinion, but because of this conflict, media often pick them up. That's why, even though there is no debate about climate change in scientific circles, you still see one being played out in the editorial pages of newspapers. And that's why television news programs still find a spokesperson with an opposite view to provide "balance" to a story—even if that opinion is patently absurd.

One could argue that it's the media who are letting us down. After all, the task of disseminating information to the public belongs squarely in their hands. Having worked as both a scientist and a journalist, I can see why that argument is tempting. But journalists work on tight deadlines and with ever-shrinking resources. Journalists with specific beats who would get to know an issue in detail are becoming scarce. And science journalists are a rare breed indeed.

Perhaps journalists could be doing a better job, but so could scientists. It isn't enough to do good work in the lab or in the field only to have your issues distorted in the press. If those who know the issues most intimately don't set the record straight, who will? In a 2005 essay in *Science,* Philippine aquaculture scientist Jurgenne Primavera made the case for scientists in the developing world to speak up, but much of what she says is universal:

> We scientists in developing countries need to come down from the ivory tower and disseminate results not only in peer-reviewed journals, but also through advocacy and the popular media. We must not forget our hearts even as we apply our minds. We do not do science in a vacuum but against the grinding poverty and environment-unfriendly character of modern times, and we can use our scientific knowledge to reduce suffering and make life more full for fellow humans and creatures.

When should scientists be advocates? Whenever they can.

THE UPS AND DOWNS OF EVOLUTION

The year 2005 wasn't an easy one for evolution, but it was a good one. In the United States, legislation to promote the teaching of "intelligent design" in schools as an alternative to evolution was introduced in more than a dozen states. But the end of the year brought court victories for evolutionists, and evolutionary research was heralded as the "breakthrough of the year" by *Science.*

Wait; didn't Darwin make that breakthrough well over a century ago? Certainly, but we must never forget that most of our understanding of biology stems from this original discovery. As geneticist Theodosius Dobzhansky once said, "Nothing in biology makes sense except in the light of evolution."

That light shone brightly in 2005. In the fall, researchers published the DNA sequence of the entire chimpanzee genome, enabling scientists to compare the genetic structure of humans with that of our closest living relatives. This research will not only help us understand human evolution but could provide important clues as to why humans are so much more susceptible than chimpanzees to problems like heart disease, AIDS, and malaria.

Other research focused on the evolutionary development of different species and how species split into two. From birds like the European blackcap to fish like the stickleback and insects like the fruit fly, researchers gained new insights into how evolution works and what causes species to stay together or become something new.

One key insight has been the increased understanding of the importance of "non-coding" DNA in speciation. This DNA does not contain instructions needed to make proteins and had no known function, so it was often labeled "junk." But we now know that the biggest genetic differences between chimpanzees and humans are found in non-coding DNA, and research into fruit flies has found that physical traits unique to certain fruit fly species can be produced in others by selectively swapping non-coding DNA.

Evolutionary research is thus vital to understanding our world. That's why scientists across the U.S. were thrilled in December 2005 when a federal judge prevented the teaching of intelligent design in Dover, Pennsylvania, biology classes. The judge reasoned that the theory, which claims that a "higher force" than evolution is responsible for the creation and development of complex organisms, is nothing more than poorly disguised creationism.

Despite the court victory, it was a challenging year for science education in the United States. As Donald Kennedy, then editor-in-chief of *Science*, wrote in 2006: "The rising tide of evangelical Christianity and its alliance with a conservative political movement seemed to foreshadow a national suspicion of science or a deep confusion about what science is or isn't."

Other criticisms were even more direct. A report by the Thomas B. Fordham Foundation in Washington, D.C., for example, warned, "Science education in America is under attack." The report gave failing grades for science education in fifteen states, including Alabama, where biology textbooks are adorned with stickers that proclaim evolution is a "controversial" theory.

Discussing intelligent design is certainly appropriate at a university level. In fact, one 2005 study published in *Bioscience* found that university students exposed to arguments for both evolution and intelligent design were actually more likely to favor evolution than those who were taught evolution alone. In other words, when it comes to advanced education, addressing belief systems rather than ignoring them could be an important teaching tool.

However, it's completely inappropriate to introduce religion into science studies at younger grades, when the capacity for critical thought has yet to develop. Canadians should be thankful that their country is by and large free of such debates. But the fact that intelligent design again reared its head so close to home means we have to be even more vigilant in ensuring that politics and religion do not cloud our teaching of science. Because when that happens, it's students who lose the most.

2

SMARTER THAN YOUR AVERAGE **PLANET**

Interconnections in the biosphere

IF **MOTHER NATURE WERE** an actor, she'd be Kevin Bacon. Famous for more than just his acting, Mr. Bacon can be connected to just about every other actor in Hollywood. But he's got nothing on nature.

Interwoven, interlinked, joined, chained, bonded, coupled—no matter how you describe it, everything in nature seems somehow connected to everything else. From the symbiotic relationships between soil microorganisms and plants to marine nitrogen isotopes that end up in cedar trees via spawning salmon that are then eaten and distributed through the forest by bears, a precious few degrees of separation may be all that lie between any two parts of the planet, no matter how geographically isolated they may seem.

That puts humanity in a bit of a spot. We've relied on our pollution "just going away" for thousands of years. As the essays in this chapter examine, that doesn't really work anymore. It's finally caught up with us. Some of the modern pollutants we create stick around for a long time and can end up in our water, our food, and our blood. Greenhouse gases like carbon dioxide from our homes, factories, and industries linger in the atmosphere for hundreds of years, continuing to warm the planet long after we've stopped emitting them. Modern technologies that allow us to fish the deepest parts of the oceans, cut down the most isolated forests, and build cities that sprawl to the horizon are also gradually winking out the life forms that have together made our planet habitable.

While many of these actions may be economically efficient in the short term, they all have repercussions. And because our knowledge of how nature works is so incomplete, many of those repercussions are unforeseen and often unwelcome. We may think that eliminating one species is just a drop in nature's ample bucket, but we usually have no idea of how that creature fits into the web of life or what will happen when the web is broken.

In some ways, we are a victim of the planet's own incredible generosity. Earth provided the ideal conditions, perhaps the only conditions, in which a species like ours might be able to evolve. We aren't exactly the most robust of creatures. We can't run terribly fast or jump very high. We lack fur to protect us from the cold. We drown if left in the water for very long. We even lack basic hunting and defensive attributes such as claws or sharp teeth.

But we do have big brains. And we used our big brains to take advantage of all that nature provided—the stable climate, the regularity of the seasons, the patterns of game migrations. We learned and took every advantage of every opportunity we could to get ahead.

So when it made sense to burn wood, then animal oils, then fossil fuels for energy, we did. And when we saw that building chimneys would send the smoke away from our homes, we built them. And when we realized that building them higher sent the pollution even farther away, we made them higher. We created a society that took full advantage of every single one of nature's services. Our success at doing this allowed our populations to soar and our economies to grow.

And when examined in isolation, each action seemed pretty small compared to the vastness of our planet. Our atmosphere seemed impossibly huge. Our oceans virtually limitless. Our soils endlessly fertile. What we didn't realize was that our air, our water, and even our soils and every living thing are all connected. These connections are what make our planet so very, very special—the only one like it in the known universe.

What happens when we ignore these connections? Included in this chapter is an essay on the Aral Sea that gives us some indication. It isn't pretty. But the story is also one of hope. Because the same interconnections that make the natural world vulnerable to human meddling also make it remarkably tenacious. And that tenacity may well be what ultimately saves us from ourselves.

. . .

LIVING WITH THE LEGACY OF "AWAY"

What do you do with waste? Why, you throw it away, of course. But think about the term "throw it away." Where exactly is this "away" place, and what happens to things when they get there?

For many North American cities, "away" could be a hundred miles or more by truck to a different county, state, or province. The city of Toronto, Canada, actually ships its trash over the American border to Michigan. And that's just household garbage being carted off to distant landfills. We also dump air pollutants up our smokestacks and out our tailpipes. We pour human and industrial waste into our rivers and oceans. We spray chemicals on our crops and hope the residue goes away.

There's an old saying: "The solution to pollution is dilution." In other words, spread the pollution out and the problem goes away. Build your smokestack higher, flush the pollutants out with more clean water, or spread them out over more land.

Of course, we now realize that this solution wasn't much of a solution at all. It just pushed the problem over to our neighbors or onto the next generation. Not long ago the world seemed like a very large place. The atmosphere seemed vast, the oceans massive. There was nothing we couldn't dump out that wouldn't go away—eventually.

Today, we are just beginning to deal with a legacy of using nature as a dumping ground. Some countries are finally starting to reduce the heat-trapping emissions that are disrupting the climate, for example, but others are still ignoring the problem. Canada and the U.S. have continued to stall on making firm commitments to reducing their emissions, which continue to rise in both countries.

Meanwhile, research published in the journal *Environmental Science & Technology* in 2005 helped show how "away" may not exist at all. On the remote west coast of British Columbia, researchers found that grizzly bears eating a diet rich in salmon are accumulating toxins that build up in the fatty flesh of the fish. These toxins, a brew of persistent organic chemicals like PCBs and flame retardants, may come from as far away as Asia, but they don't just disappear. They can continue to exist for decades and be transported halfway around the world.

Researchers say that they don't know what effect these toxins are having on the bears, but the chemicals are known to mimic animal hormones and may pose a developmental risk to young cubs. The same chemicals are turning up in much higher concentrations in killer whales and polar bears.

Humans aren't immune to our legacy of hoping things go away either. Bulging landfills, smoggy skies, and a disrupted climate are some of the most obvious problems we face as a result. But this legacy often affects us in ways we can't even see. For example, in 2005, another paper published in *Environmental Science & Technology* reported that rice grown in the United States may contain up to five times more arsenic than rice grown in Europe or Asia.

In this case, the culprit was thought to be arsenic-based pesticides sprayed on cotton fields throughout the southern U.S. Many of those fields have since been converted to rice crops, and arsenic left in the soil is finding its way into the grains. In 2008, researchers with Cornell University updated the study, confirming that the American rice contained higher levels of arsenic. Fortunately for human health, it is in the form of methylated arsenic, which is considered less toxic than the inorganic form of the metal.

Still, such examples point to the ever-decreasing distance between "away" and human beings. We live in a disposable culture where it's easy to forget that things we throw away don't necessarily go away. And unless we start ensuring that the chemicals and other junk we release into the environment readily break down and don't build up over time, we will continue to build on our legacy. "Away" may not be on a map, but it's now closer than ever.

*   *   *

SPECIES LOSS WEAKENS ENTIRE ECOSYSTEMS

Each year, the International Union for Conservation of Nature (IUCN) releases its annual Red List of Threatened Species. In 2007, another two hundred species were added to the existing list of more than sixteen thousand. And in 2008 the IUCN released its most comprehensive assessment of mammals ever done, which revealed that one in four is at risk of extinction.

Many people often don't realize that the decline of these species isn't just a sad story that's happening "out there" in nature; it's really a story that's happening to us, something that we're doing to ourselves. And it's making all of the eco-systems that we ultimately depend on biologically poorer and more vulnerable.

Living in cities, it's easy to forget how much we depend on the services provided by healthy, natural ecosystems—things like cleansing water, filtering air, and storing carbon to reduce global warming. Our health and well-being depend on these services, which have also been conservatively estimated at being worth trillions of dollars to the global economy.

However, reading stories about how species are being pushed to the brink of extinction doesn't necessarily trigger alarm bells about our own future. Many of the animals in these types of stories have exotic names from faraway places, like the Yangtze River dolphin and the western lowland gorilla, so it's easy to gloss it over as someone else's problem. But the reality is, in an interconnected world their problem is our problem.

As hard as it may be for some people to believe, the other species of the world don't exist just to look pretty and give tourists something to photograph. They actually occupy ecological niches. Their mere existence is often vital to the overall health of the ecosystem. Losing a species or having one pushed to the brink of extinction can have what biologists call "cascading" effects on the entire region.

Consider the role of large primates in tropical forests. In these forests, large primates play several important roles, one of which is in seed dispersal. Many tropical primates are frugivores; that is, their diet consists largely of fruit. While small-seeded fruit trees may have a large number of species, including mammals, reptiles, and birds, to help them spread their seeds, large-seeded tropical fruit trees rely largely on bigger mammals—especially primates.

When primates like monkeys, apes, and chimps eat fruit, they physically spread the seeds over a wide area of forest floor. So the animals receive sustenance from the fruit while the trees get their seeds spread across a large area, allowing them to grow elsewhere, which then provides more food for the primates. It's a mutually beneficial relationship.

But when large primates are hunted to greatly reduced numbers, as they increasingly are, it can have a profound impact on the ecosystem. For example, in 2007, a special edition of the journal *Biotropica* focused on the impact of what's called the "bushmeat" trade—local hunting that often includes primates. In one study, researchers from the University of Illinois looked at two sections of Peruvian forest. One section had been heavily hunted by local people using modern weapons, like shotguns, and had lost more than 80 percent of its large primates. The other section was protected from hunting. The researchers found that there were 55 percent fewer species of large-seeded fruit trees in the unprotected forest and 60 percent less of the fruit trees themselves. In other words, once the large primates were gone, the trees that depended on them started to disappear too.

Of course, as the researchers point out, this has a number of unfortunate consequences. It makes the forest less hospitable to large primates, so they are less likely to be able to ever come back. Having less fruit tree diversity makes remaining primates more vulnerable in times of scarcity. And the trees themselves, which often have economic value to humans for timber, fruit, or other uses, start to disappear.

Humans depend on the services provided by healthy eco-systems, so it's in our best interests to conserve the creatures that live in them. Losing one species or having two hundred more hunted off to the brink of extinction isn't just sad news for us—it's dangerous.

.   .   .

SAVING ROY'S LAKE

What does nature mean to you? For some people, their land defines who they are as a culture. For others, it provides a place of escape and renewal from the frenetic urban centers where the majority of North Americans live. But no matter how we feel about it, nature's future is increasingly dependent on the decisions we humans will collectively make over the coming years.

A couple of years ago Roy MacGregor, columnist for one of Canada's major newspapers, *The Globe and Mail,* wrote a lovely article about spending time on the lake and in nature: "If you head down the lake and through a small narrows—watch out for that deadhead—you come to a smaller lake with not a single dwelling to be found. The surrounding leaves hint at what time of year it is, but says nothing of the year itself. It could be this century, it could be last century—could, with an enormous amount of good fortune—be next century."

It's a beautiful and evocative passage, and one that I carry with me always. But I'd make one small alteration to the last sentence: "It could be this century, it could be last cen-tury—could, with an enormous amount of good fortune *and effort*—be next century."

You see, it's not just about good fortune anymore. We aren't just subject to the whims of nature. We're subject to the whims and pulse of humanity. Whether or not Roy's lake will be around next century depends far less on good fortune and far more on what we as a society and what we humans as a species decide to do in the next few decades.

That lake may look the same as it did a century ago, but it's actually probably undergone some fairly substantial changes. The water no doubt contains persistent, man-made chemicals that didn't exist a century ago. The pH level of the water has probably changed because of acid rain, and the wildlife around the lake has changed in abundance and diversity as a result of hunting, trapping, and logging in the surrounding area. More changes, too, are underway due to global warming, and these could be the most significant yet.

Still, despite the changes that have taken place, North America is still blessed with some of the largest intact healthy ecosystems in the world. It has a huge land base and a relatively small population compared to Europe or Asia. But its citizens are just not taking very good care of it.

By most measures, North Americans are failing. While some individual states, provinces, and cities are indeed making strides towards environmental sustainability, the federal governments of Canada and the United States have failed miserably. Most other industrialized countries are at least trying to deal with climate change, air and water pollution, and species extinction. The U.S. and Canada are not. Other countries are working towards becoming sustainable. North American countries are not.

That's a hard fact for some folks to face when they're camping in the great outdoors, sitting on the dock of a lake, or canoeing down one of North America's countless magnificent waterways—but it's true. It's only by an enormous amount of good fortune that there is still an opportunity to change before it's too late.

Conserving the natural world that many of us hold so dear is up to all citizens, but few of us have the power that our elected leaders have. They make the laws and they make sure the laws are enforced. These are the folks we need to talk to if we want to make change happen. Of course, that's up to us too. Unless we ask, they won't do a thing.

So, when you leave the great outdoors—whether it's a favorite fishing hole, a lakeside cabin, a national park, or even a greenbelt near the city—and head back to the hustle and bustle of your everyday life, try to take a little bit of those special places with you. Places like Roy's lake, where we truly feel connected to the land. Then remember that it's up to us to make sure these places are still there in the next century, so our children and grandchildren can enjoy them too.

.   .   .

HUMAN HORMONES MESS WITH MALE FISH

Most people alive today were born after 1950. To these people, our modern world is just the way things have always been. Imagining life without TV, radio, telephones, and the internet is next to impossible. Teenagers probably have a hard time imagining life without text messaging!

And it's true: human reach is now profound. We are the most integrated, interconnected, and mobile species that has ever existed on this planet. Some of these interconnections produce marvelous results. We get to know other cultures. We understand more about history and each other. We can easily chat with friends and family on the other side of the world.

But we have to remember that, although we are connected with each other more than ever, we are also intimately connected to the rest of the natural world. These connections can manifest themselves physically, such as through global warming. But they can also manifest themselves biologically—and in some surprising ways.

In 2007 researchers writing in the U.S. journal *Proceedings of the National Academy of Sciences* reported that male fish became "feminized" when exposed to human hormones. Some of the fish, a type of fathead minnow, produced early-stage eggs in their testes, while others actually developed tissues for both reproductive organs.

How would fish be exposed to female human hormones? Through treated or untreated municipal wastewater, of course. It seems that widespread use of birth control pills has elevated the amount of estrogenic substances going into our waste stream. Remember, things that go down our toilets don't just disappear. They can actually survive simple sewage treatment processes and end up in our rivers, lakes, and oceans.

Reports of fish feminization as a result of human female hormones are fairly well documented—but no one had done long-term studies of the impact this can have on fish populations. For this study, researchers actually added the synthetic estrogen found in contraceptive pills to a remote lake in northern Ontario in amounts normally found in human wastewater. They did this for three years and monitored the results over a period of seven years.

The results were startling. As expected, the male fish developed some feminized characteristics, such as producing proteins normally synthesized in females. But what really disturbed the scientists: populations of the fish crashed to near extinction levels by the end of the experiment. Feminization of the males combined with hormonal changes in the females apparently damaged their overall reproductive capacity to the point that the fish were unable to maintain their population.

The researchers conclude: "The results from this whole-lake experiment demonstrate that continued inputs of natural and synthetic estrogens and estrogen mimics to the aquatic environment in municipal wastewaters could decrease the reproductive success and sustainability of fish populations."

This spells trouble. Most of us have probably never heard of the fathead minnow, but these fish are a vital food source for well known and popular sport fish that people have heard of—such as walleye, lake trout, and northern pike. They are also well-studied and often used in toxicology testing because they have short life cycles, adapt well to lab conditions, and are representative of a large family of fish.

The authors of the report describe the fathead minnow as "a freshwater equivalent of the miner's canary." In other words, what happens to the fish, as with the bird, could happen to humans in short order unless we are very careful. Cell phones and the internet aren't our only connections with each other and with the world. We are biological creatures too, and we have to remember that our biological connections ultimately matter the most.

.   .   .

OCEAN LIFE MAKES WAVES

Most people have heard of the "butterfly effect"—the idea that a small change, such as a butterfly flapping its wings in one part of the world, can set in motion a series of events that leads to a big event, such as a tornado, somewhere else. The term is largely used as a metaphor, but science now shows that there's a literal aspect to the theory that has much broader implications.

To say that everything is connected to everything else has become a cliché, but it's true—especially in nature. Scientist and author James Lovelock uses the term "Gaia" to describe the Earth as a living, self-regulating system. According to this hypothesis, all of the planet's biological creatures are intimately connected with all of its physical systems, from the soils to the oceans and the atmosphere. Changes in any of these systems can affect everything else.

We can see how connected everything is when we release long-lasting substances into the atmosphere. Toxins, for example, can drift out of a smokestack in Hamilton, Ontario, or Mumbai, India, circle the Earth on the winds or ocean currents, and end up in seemingly pristine areas such as the Far North. In fact these toxins are now found concentrated in the fat of marine mammals and in human breast milk.

In an interconnected world, even a creature as small and seemingly inconsequential as the tiny, shrimp-like krill can have a big impact. These half-to-three-quarter-inch-long creatures already play an important role in the ocean food chain and are a staple in the diet of some of the world's largest whales. But krill are so small that few people would have suspected they could play an important role in generating currents that help mix our ocean waters.

Yet that's exactly what a team of researchers from the University of Victoria found off the coast of British Columbia's Vancouver Island. The researchers looked at a deep ocean inlet where different layers of the water mix very little. They found that millions of krill, on their nightly upward migration from the deep water to the surface to feed, increased the mixing of water by three to four orders of magnitude. In other words, these tiny creatures actually cause quite a stir.

And it isn't just krill that cause this water-stirring or "turbulence." All living organisms that exhibit similar behavior can cause turbulence, helping to bring cold, nutrient-rich water up to the surface. This exchange of cold and warm water is vital to the productivity of the oceans. It also helps break down human wastes, and it even plays an important role in the climate.

But turbulence has largely been thought to be driven almost exclusively by physical forces like the winds and the tides rather than by biological forces. The very idea that the behavior of individual organisms can affect entire systems seems fantastic. Yet the researchers in Victoria concluded that sea creatures themselves may be a critical, but overlooked, source of turbulence in the oceans.

Other researchers go even further. A 2006 article in the *Journal of Marine Research* calculated that, based on the math, swimming organisms may be one of the most important drivers of ocean turbulence. If this is the case, the authors concluded, then the overfishing that has caused fish stocks to plummet and the near extinction of many whale species as a result of hunting may have disrupted ocean turbulence enough to affect the planet's climate.

Seemingly small actions can have big consequences. More and more, we are finding that our world is not nearly as vast and limitless as we once supposed. Not only is it interconnected, but this very interconnectedness drives it. In this world, we are all butterflies, and we need to be mindful of what can happen when we flap our wings.

VAMPIRES ON THE LEADING EDGE

"Rabid vampire bats attack Brazilian children" may sound like something out of the tabloid *Weekly World News*, but the headline actually comes from a 2005 article published in the respected magazine *New Scientist*. Even weirder—it's true.

Vampire bats had indeed been attacking Brazilian children.

In fact, they bit more than thirteen hundred people in Brazil in 2005, and twenty-three of their victims died from rabies. But beneath this sensational and bizarre story is more hopeful news about the emerging field of conservation medicine.

Conservation medicine is a relatively new discipline referring to the convergence of ecology, which looks at species and ecosystems, and health science, which looks at human, plant, and animal health. It's a natural connection because the health of individual plants, animals, and people is intimately connected to the health of the ecosystems in which they are embedded.

What does this have to do with bats? Well, since 2004, the main transmission source for rabies in South America has shifted from dogs to vampire bats. And the primary reason for the increase in vampire bat attacks in Brazil and other areas of South America is thought to be deforestation. The Amazon forests are being rapidly cleared for industry and agriculture—especially grazing animals. With their homes gone, the bats are roosting closer to humans and they have a new, plentiful supply of slow-moving, warm-blooded prey—cattle. This has led to larger colonies in smaller areas, ideal breeding grounds for rabies, which makes the bats more aggressive and no longer fearful humans. In 2006, a rabid vampire bat was found in an urban area of Brazil, the first such case ever recorded.

And rabies isn't the only disease recently transferred to humans from bats. Bats are also a natural reservoir for SARS, the respiratory virus that caused panic in Toronto and tore through Southeast Asia a few years ago. Originally, scientists thought civet cats were the reservoir for SARS, but they now believe the civets were infected by bats. Bats often don't eat all of their meals. Fruit bats, for example, chew fruit to extract the sugars and then spit out the pulp—which is eaten by foraging animals below.

Scientists now believe that this is how the Nipah virus was spread through pig farms in Malaysia several years ago, when farms began displacing forests and bats began roosting in barns. Authorities there had to kill one million pigs, and more than one hundred farm workers died from the virus. More recently, in Bangladesh, the Nipah virus was spread directly to humans when children picked and sold fruit that had been contaminated with partially digested bat dinners. Researchers also believe fruit dropped by bats may have spread the Hendra virus in Australia and the Ebola virus among primates in Africa.

But before we get out the pitchforks and torches to hunt down these winged terrors, consider what ecologist Andrew Dobson wrote in an analysis in *Science*: "Assuming we can control these diseases by simply controlling bats is both naïve and short sighted. Instead, we must recognize that increased spillover-mediated pathogen transmission from bats to humans may simply reflect an increase in their contact through anthropogenic modification of the bat's natural environment."

In other words, as humans continue to modify and destroy bat habitat, we will continue to run into these problems. To solve them, we must focus on conservation and learning more about bat ecology and immunology—about which we currently know very little. Ultimately, minimizing the conditions that lead to disease outbreak is much more effective than dealing with the problem after it has already occurred.

In nature, everything is connected. And while people tend to think that human society is somehow excluded from nature like some sort of observer, we are in fact deeply embedded in the natural world. Because of this, our actions can have profound, unforeseen, and mysterious repercussions. The new field of conservation medicine can help unlock those mysteries and help us build a healthier world.

MESSING UP A PLANET IS EASY; FIXING IT TAKES TIME

Most of us are used to life moving at breakneck speed. We carry cell phones so we can be reached instantly. We use email so we can transmit text and photos in the blink of an eye. We eat at fast-food chains so we can get our food immediately. We drive everywhere to get there faster. Once we've made a decision, we want results—now.

Unfortunately, the rest of life on our planet doesn't work that way. Things take time. Processes evolve over hundreds, thousands, or millions of years. As a result, humanity has no problem messing things up quickly, but little patience when it comes to fixing them.

A perfect example is the ozone layer. Over twenty years ago, scientists discovered a massive "hole" in the ozone layer, the protective layer of ozone gas high in the atmosphere that helps shield all living things from the sun's harsh rays. The hole was growing and threatened to cause increased skin cancer in humans and a host of unknown environmental problems.

Over the next few years, scientists determined the cause of the problem—a group of chemical compounds called CFCs, which were used in solvents and aerosols and as coolants in refrigeration units and air conditioners. When CFCs find their way up into the stratosphere and react with ultraviolet light, it creates chlorine free radicals, which are potent scavengers of ozone. In an unprecedented move, nations around the world quickly agreed to phase out CFCs and, in 1989, the Montreal Protocol was born.

The Montreal Protocol is a well-known and unqualified success. As one researcher pointed out in an article in *Nature*, even schoolchildren today are familiar with the story. Unfortunately, our instant-fix mentality is so ingrained that many people are still confused when stories about the ozone hole continue to appear every year. "Didn't we fix that?" is a common refrain.

In fact, the Montreal Protocol is working. CFC production has dropped to near zero, and the ozone layer seems to be gradually repairing itself. But CFCs can persist in the atmosphere for fifty to a hundred years. So some of the CFCs manufactured forty years ago are still destroying ozone today. It will take decades before the protective layer fully heals. Until then, the size of the hole will fluctuate from year to year.

Another class of chemicals that will continue to haunt us for decades, even though a number of them were banned in 2001, is persistent organic pollutants. These toxic chemical compounds, which include PCBs, DDT, and dioxins, are easily transported by air and ocean currents and have found their way into even the most remote regions of the planet. They did not exist eighty years ago, but today traces of these compounds can be found in the bodies of every person on Earth. No one knows what long-term effect these substances are having on our health.

As humanity's influence on the environment and natural systems continues to grow, we have to remember that it can take far longer to solve our problems than it does to create them. We cannot just switch a problem off like a remote-control television. Global warming, for example, will not be solved instantly. The carbon dioxide we pump into the atmosphere today will stay in our air for several hundred years. Even if we stopped producing heat-trapping gases today, the Earth will continue to warm, and we will continue to have more extreme weather events and other climate-related problems for generations.

That's why it's so important to get started now. Our planet cannot be commanded to fix itself. Mother Nature does not have a cell phone. She doesn't use email, and she's not too keen on instant messaging either. She takes her time, and we'd better get used to it because like it or not, we're on her schedule.

. . .

THE DISAPPEARING SEA

Khiva, Bukhara, Samarkand, and Tashkent, once oases along the fabled Silk Road to Cathay, are names that still conjure up images of a time long past. Today they are cities in Uzbekistan, one of five independent republics in Central Asia that formed after the breakup of the Soviet Union. In 2000 I spent two weeks there, chronicling an environmental catastrophe that shows how intimately connected our health and economies are to our environment. If ever there was a cautionary tale about how quickly and efficiently humans can destroy an ecosystem, this is it.

Millions of years ago, the northwestern part of Uzbekistan and southern Kazakhstan was covered by a massive inland sea. When the waters receded, they left a broad plain of highly saline soil. One of the remnants of the ancient sea was the Aral Sea, the fourth-largest inland body of water in the world. The Aral Sea was a rich source of fish. Biologists identified some twenty species, including sturgeon and catfish. The town of Muynak, located on the edge of the sea, was a fishing village that also attracted tourists to its seaside vistas.

In the 1950s, the Soviet Union decided the great plains were ideal for growing cotton. The critical factor to make it happen was water. Two great rivers feed the Aral Sea: the Amu Darya and Syr Darya. The Soviet scheme was based on the construction of a series of dams on the two rivers to create reservoirs from which nearly 25,000 miles of canals would eventually be dug to divert water to the fields. The fields flourished, but, with such vast areas of monoculture, farmers had to use massive amounts of chemical pesticides. And with irrigation, salt was drawn to, and accumulated on, the soil's surface. When the Tahaitash Dam was built on the Amu Darya River near the city of Nukus, no water remained in the riverbed to flow to the Aral Sea, hundreds of miles away.

To the surprise of the inhabitants of Muynak, the Aral Sea began to shrink. At first, they assumed it was a temporary condition and dredged a canal to the receding shore so boats could continue to ply the sea and dock at the wharves. But the effluents that did reach the sea were laced with a deadly mix of salt and pesticides from the cotton fields. Fish populations plummeted, and eventually, when the canal was nineteen miles long and the sea continued to move away, the boats were abandoned to lie like great leviathans on sands that were once sea bottom.

By the time of my visit, Muynak was a desert town more than sixty miles from the sea. The only reminders of the once thriving fishing activity were the rusting hulks of ships and an ancient fish plant. The sea had shrunk to two-fifths of its original size and ranked about tenth in the world. The water level had dropped by fifty-two feet, and the volume was reduced by 75 percent, a loss equivalent to the water in both lakes Erie and Huron. The ecological effect had been disastrous and the economic, social, and medical problems for people in the region catastrophic.

All twenty known fish species in the Aral Sea were extinct, unable to survive the toxic, salty sludge. In just a few decades, a centuries-old way of life disappeared. The vast area of exposed seabed was laced with pesticides, so when the wind blew, dust storms spread salt and toxics. It's estimated that eighty-three million tons of toxic dust and salts are spread across Central Asia each year. If the Aral Sea dries up completely, seventeen billion tons of salt will be left behind.

I was shocked at how these pollutants had taken their toll on local residents. People in the area had the highest incidence of tuberculosis in the world, as well as elevated levels of bronchial and kidney problems and cancer. Leonid Elpiner of the Russian Academy of Sciences recited a litany of problems to me, ranging from intestinal diseases, polio, viral hepatitis, and noninfectious diseases to higher levels of chemical pollution of air, water, and food.

Zita Mazhitova, head of pediatrics at Kazakh State University, told me that 80 percent of fertile women in the region were anemic and 87 percent had various chronic diseases. Mortality had doubled and life expectancy dropped. Children were especially vulnerable.

"Each child from the Aral Sea area has many affected organs and systems at the same time," she said. "Half of the children suffer from insufficient weight and retarded growth and puberty. All children examined had immunodeficiency to various degrees . . . All the children suffer from disease of ear, nose, and throat; 83 percent have skin and mucosa lesions . . . every second child has chronic bronchitis . . . There is a high frequency of inborn defects of respiratory organs and bronchiectasis. The latter is caused by [the] impact on mucosa of [the] respiratory tract by [the] toxic mixture of fine-grained salines, pesticides, and dust lifted by winds and storms into the air and transferred for long distances."

Mazhitova's devastating analysis concluded: "There are no healthy children in the Aral Sea area and 89 percent of them have several organs and systems affected at the same time."

By 2000, an estimated one hundred thousand environmental refugees had already left the region, forty-two thousand in Kazakhstan alone. But the impact of the Aral Sea disaster had spread far beyond the adjacent communities. Storms blew over hundreds, probably thousands, of miles. And the region's climate had actually changed because of the loss of the sea's moderating effect. Winters were colder and summers hotter— making dust storms even worse.

It would be a mistake to conclude that the only culprit is solely the system of Soviet-style central planning. As Dr. Viktor Duchovny, formerly of the Ministry of Water, told me, mega-projects to develop and use water were pushed extensively in the United States and Canada—although without such tragic consequences, so far.

The real lesson: it is a mistake to think we are so clever that we can shoehorn nature to fit our agenda. We depend on complex rhythms and cycles of nature—seasons, climate, soil hydrology, flora, fauna—all of which we barely understand. We have acquired the brute power of technology, but we have applied it without proper respect for the interconnectedness of living systems or the recognition that we don't comprehend all of nature's details.

The Aral Sea is a fable for our times, a warning. We have had our own lessons, but so far not on this great a scale. We know Love Canal, Chernobyl, and the Sydney Tar Ponds. And if we think about it, the way we are treating our own wetlands, boreal forests, old-growth temperate rainforests, and rivers represents the same mindset that was responsible for the Aral Sea disaster.

But just as much as the Aral Sea is a disaster story, it is also one of hope. Our little planet never ceases to amaze me. We keep kicking the stuffing out of her and she keeps finding a way to bounce back.

By 2007, the Aral Sea had shrunk to just 10 percent of its original size and split into three isolated lakes. But the North Aral Sea is showing signs of improvement, thanks to a rescue

mission funded through the World Bank. After years of reha-
bilitation, including the construction of new dams, dykes, and
sluices to repair previous damage, the Syr Darya River now
flows freely and the amount of water reaching the sea has
doubled. In fact, since 2005, the northern sea level has now
risen by more than twenty-five feet.

For the North Aral's main port city, Aralsk, the project has
been a lifesaver. When sea levels started to fall decades ago,
more than half of the population left, and eventually residents
found themselves fifty miles from the water's edge. Today, the
water has risen to within nine miles of the city, and fish have
even started to return, providing a source of food and income
once more. Rising sea levels have also increased rainfall levels
and reduced the number of toxic dust storms that plagued the
region.

Sadly, the future of the South Aral Sea is not as rosy. The
South Aral is largely in the poor country of Uzbekistan, which
continues to draw heavy amounts of water to irrigate cotton
fields. Although the World Bank is working on a project to
restore some wetlands in the South Aral, the sea continues to
shrink and, according to some estimates, could dry up completely in the next ten years.

It will be a very long time, if ever, before we can say the
Aral has truly returned to its former glory. But the recent
improvements to the North Aral Sea show that if we make
an effort to change our ways, the planet is capable of healing
itself. With enough political will, we can move mountains.

Does this mean we don't really have to worry about environmental problems since we can just fix them later? Not at
all. Cleaning up our messes after the fact takes an enormous
toll on human health and finances. The World Bank has spent
US$100 million so far to partially restore the northernmost,
and smallest, of the three lakes that now make up the Aral

Sea. In addition to the financial cost and direct human health impacts of poor environmental planning, some of the changes we're making to the planet will affect us for centuries. Greenhouse gases like carbon dioxide, for example, can stay in the atmosphere for hundreds of years. Even if we stop churning them out today, the climate will continue to heat up.

With some issues, like climate change, we simply cannot afford to take a wait-and-see approach. The longer we wait, the more expensive and difficult the problem will be to fix later on—if it's even possible to fix. In light of all that we have learned from the Aral Sea disaster, experimenting with the only atmosphere we have seems unbelievably foolish.

Our planet may have a remarkable capacity to heal, but there's only so much she can take.

3

GETTING TO KNOW THE **JONESES**

Protecting the diversity of life on Earth

NEIGHBORS. THEY CAN BE friendly, good-natured folk who mind their own business and occasionally drop by for a chat—sometimes bearing a nice, warm bumbleberry pie. Or they can be beer-swilling yahoos who rev their Harleys at 2:00 AM while blaring Lynyrd Skynyrd tunes.

I've had both. But more often than not, these days most people don't really know their neighbors very well. Or even at all.

That's a shame, because social scientists tell us that it's these kinds of social networks that we really rely on in times of need. Think of the way your neighbor might help out if your house was on fire, for example. What's more, human interaction tends to make us happy. Getting to know our neighbors and our communities is generally considered healthy and beneficial—both for the individual and for the community itself.

The same could be said of nature, only on a much larger scale. Human beings are but one part of a massive community of life on this planet—a community about which we know surprisingly little. Yet it's a community that benefits us greatly.

As we discuss in this series of essays, our neighbors are all around us. And they come in all shapes and sizes, from tiny microscopic organisms that we really don't understand to great big charismatic carnivores like the polar bear—that we don't really understand very well either. Each of these creatures has a role to play in its ecosystem, and each ecosystem plays a part in the functioning of the Earth's biosphere.

What it comes down to is this: diversity of life on Earth at all levels is very important. It's important from both an evolutionary perspective and from a stability perspective. A diversity of life on our planet is a big part of what keeps the entire system functioning. Studies have shown that the more diverse life is in an ecosystem, the more stable the ecosystem is and the less vulnerable to change. As our climate changes, more diverse ecosystems are more likely to be able to better adapt.

But many ecologists say that we are also in the middle of a human-induced extinction event. Life forms are believed to be winking out all over the planet—often without us ever having acknowledged their existence in the first place. Sometimes it's a result of hunting and fishing. Sometimes it's from habitat destruction. And sometimes it's from pollution. Really, it's a result of all of these and a thousand other forces acting separately and together to put enormous pressures on our neighboring creatures.

All of this is very complicated and very depressing. But despite our rather destructive tendencies, I believe that humans really do care deeply about the other creatures on our planet—an innate caring that Harvard ecologist E.O. Wilson called "biophilia." We connect to our planet through other wild creatures. It's how we relate to the otherwise often incomprehensible complexity of the natural world. We personalize it by looking at the individual parts, the individual species. We want to know their stories. And the more like us they are, the better we will understand them.

So by all means, let's learn those stories. Let's get to know our neighbors. Because more and more we're finding that they aren't just yahoos revving their Harleys at two in the morning; they're actually helping us in countless ways. They're helping to keep all the services that we need functioning. They bring stability to our ecosystems—ecosystems we depend on to clean our air and water, provide food, and store carbon to slow global warming. They aren't just pitching in to douse the flames when our house is on fire; they're actively working to make sure it keeps standing all the time. They don't just *bring* bumbleberry pies; they *are* bumbleberry pies.

And I don't know about you, but those are the kinds of folks I'd like to meet.

. . .

UNDERSTANDING DIVERSITY
THE FIRST STEP IN PROTECTING IT

A massive 2004 deep-sea expedition in the North Atlantic uncovered hundreds of species of fish and squid, including several that appeared to be new to science. The expedition made headlines, but some readers may have been left wondering, "So what?"

The two-month international expedition netted some eighty thousand specimens from waters up to several miles deep. Scientists were thrilled with their discoveries, which included several potentially new species along with a variety of strange phenomena, from reef-building cold-water coral to planktonic organisms arranged in rings more than six miles wide.

Like a comparable study in the Pacific Ocean, it's all part of the ten-year Census of Marine Life. But at the cost of more than US$1 billion, some might say the price seems a bit high to just find a few new fish. So what makes this kind of work so important?

Well, beyond the simple thrill of gaining knowledge for its own sake, understanding life's diversity on the planet and how it interacts is vital to humanity. After all, how can we claim to "manage" wild organisms if we don't even know what they are? In nature, diversity rules. A diversity of life has proven to be a key element of evolution and the resilience of life on Earth over long periods of time, even through periods of great change. The more we understand about diversity in nature and how to protect it, the better off we will be.

Faced with a growing population of some 6.7 billion people and an explosive rise in the demand for energy and natural resources, the planet's diversity of life is under assault at all levels—from the number of species on the planet (species diversity) to the number of different populations of a species within an ecosystem (population diversity) and the variation within populations (genetic diversity). Although most of our focus tends to be on species diversity, population and genetic diversity are also extremely important to the health of an ecosystem.

A 2004 study out of the University of Toronto, for example, found that genetic diversity within a plant species can have the same effect as species diversity when it comes to influencing the variety of life in an ecological community. For the study, researchers planted clusters of a species of evening primrose at field sites, some of which were a monoculture of one genetic variation, others of which contained a number of genetic variations. They found that the most diverse clusters attracted 17 percent more species of insects, spiders, and other arthropods.

Again, however, the skeptic might ask, "Great, so greater genetic diversity means more bugs. So what?"

Well, on a large scale, this diversity of life on Earth underpins many of the ecosystem services we need to survive. Microorganisms in our soil, for example, help make it fertile, while microorganisms in our oceans provide us with the oxygen we need to breathe. For humans, these services are essential to our very survival, so they are priceless. Preventing the degradation of these services is therefore essential.

Of course, humans don't set out to deliberately harm ecosystem services. Instead, we do harm on an incremental scale, site by site, ecosystem by ecosystem. That makes the impacts less obvious—especially when we don't understand how species and processes interact with one another in the first place.

Fortunately, some ecosystem services take place on a scale small enough to measure with a dollar value. A study in Costa Rica found that preserving fragments of original forest around coffee plantations could boost crop yields and increase income. Researchers found that the forest offered a refuge for bee species, which helped pollinate coffee plants. Plants closer to the forest received more pollen more often from these bees than plants farther away. As a result, plants near the forest yielded 20 percent more beans.

So what's the big deal about finding fish in the middle of the ocean? It may not have been in the news stories, but it's all part of the big puzzle of understanding the diversity of life on Earth, how we benefit from it, and how we can protect it. And that is a pretty big deal indeed.

LISTEN TO THE BIRDS

Most everyone has heard the analogy of the canary in the coal mine. Miners used to take the birds, which are hypersensitive to deadly hydrogen sulfide gas, down into the shafts. If the bird keeled over, miners knew to get out—fast.

Today, birds are warning us once again, only on a much larger scale. A 2004 report from the conservation group Bird-Life International found that one in eight of the world's bird species is facing extinction and one-third are at risk. It's the first time that one paper brought together status reports of bird populations worldwide for a true global analysis.

To this day, it remains the single most comprehensive global analysis of bird life, and the findings were pretty grim. More than 1,200 bird species face extinction, with some 200 on the critical list. Around the world, from the Canadian prairies to Africa and India, bird populations are in trouble—largely because human activities are damaging their habitats. Converting prairie grassland to farmland, for example, has resulted in a 60 percent decline in native prairie bird species. Similarly, in Africa, 50 percent of all birds are threatened by agricultural expansion. And unsustainable forestry practices in the tropics threaten hundreds more species.

Some birds are threatened in seemingly bizarre and unexpected ways. Longline fishing, for example, kills tens of thousands of albatrosses every year. The birds die when they swallow freshly baited hooks, which then sink, pulling the birds down with them. Literally hundreds of millions of hooks are baited on longlines, and albatrosses have come to see fishing boats as a food source—albeit a very dangerous one. Seven species of albatross are endangered and three are critically endangered because of longline fishing.

In India, three vulture species face imminent extinction from eating livestock carcasses tainted with diclofenac, an

anti-inflammatory drug. Diclofenac is widely used to treat livestock, and the drug remains in an animal's body for days. If the
animal dies during this time, vultures are such good scavengers that it is not uncommon for hundreds of vultures to find
and feast on a single carcass. Unfortunately, vultures are so
sensitive to the drug that most will then die. In fact, 85 percent of vulture deaths in India can be traced back to this drug,
which has resulted in a 95 percent population decline of these
species in just eight years. BirdLife International calls this the
fastest bird species extinction in 150 years—the worst since
the demise of the passenger pigeon.

Although India has restricted the sale of diclofenac and is
trying to rehabilitate vulture populations, the birds are still in
trouble because many farmers continue to use imported Chinese versions of the drug for their livestock. The loss of vulture
populations as scavengers has resulted in an increase in fes-
tering animal carcasses, which has led to a massive increase
in feral dog populations. Between 1992 and 2006, the loss
of the vultures is estimated to have allowed the Indian feral
dog population to increase by more than five million animals.
This has resulted in an increasing number of dog bites, which
has resulted in more cases of rabies in humans. India now has
the highest incidence of rabies in the world. Each year, more
than twenty thousand people in India die from the disease.
Developing countries such as India face the greatest pressures
yet have the fewest resources to protect birds. Nigeria, for
example, which is slightly larger than Texas, has more than six
times the state's population and a mere fraction of the wealth.
Conserving bird species in such a land, where people are
more concerned with finding food and shelter, is extremely
challenging indeed.

Despite North America's natural and economic wealth,
however, we aren't doing a good job of protecting our birds

either. The Endangered Species Act in the United States has helped several bird species recover, but more than two hundred species are still endangered, many critically. In Canada, federal endangered species legislation has shown little success. With some two dozen bird species in Canada endangered, the legislation had better prove its worth soon because for some birds, like the spotted owl, there isn't much time left. Many people might remember the spotted owl as the poster child for forest conservation in the United States, but the owl is actually far worse off in Canada. Just seven Canadian birds are known to be left in the wild, but logging in their habitat is still allowed.

We must hope that Canada's Species at Risk Act shows some teeth—as much for ourselves as for the birds. Bird populations are widespread and well studied. Their health tells us something about the environment as a whole, and if it suffers, we all do. "Canary in a coal mine" was originally a literal reference that became a metaphor for a warning signal. Today the coal mine is the Earth, and we have nowhere to run.

.   .   .

BUTTERFLIES MAY MAKE BETTER CANARIES

As important as birds are as an indicator species to help us judge the overall health of life on planet Earth, there may be an even better canary in this big blue coal mine—butterflies. In 2004, results of a long-term study of bird, butterfly, and plant populations in Great Britain were published in *Science*. The results confirmed that, in Great Britain at least, many bird species are indeed declining. But even more disturbing, the researchers found that birds and other vertebrates may not necessarily make good indicator species because lesser-known creatures like insects seem to be faring much worse.

For the Great Britain study, researchers looked through fifteen million records of species amassed by some twenty thousand volunteers. Over forty years, these volunteers kept detailed records of more than three thousand separate ten-square-kilometer (four-square-mile) test areas across the country. Great Britain is the only place in the world where such detailed records have been maintained for so long.

After studying the data for a year, researchers concluded that 28 percent of plant species and 50 percent of bird species had disappeared from at least one study area. Butterflies fared the worst, with 71 percent of those species disappearing from at least one area over the past twenty years. In fact, two butterfly species went extinct from Britain during the study period, as did six native plants.

These findings are disturbing because insects account for more than half of the known species on the planet. According to the researchers, if insects are disappearing faster than birds, then biologists have actually been underestimating the loss of life on Earth. The findings thus "[strengthen] the hypothesis that the natural world is experiencing the sixth major extinction event in its history."

While the last major extinction event occurred when an asteroid collided with the planet some sixty million years ago, current extinction levels have a much more mundane cause—human activities. Sometimes the effects of these activities are obvious—like when we fill in a wetland to build a parking lot. Others are more subtle. For example, results of a study published in *Science* found that excess nitrogen from intensive agricultural production and air pollution in Great Britain is reducing plant diversity by creating conditions more favorable to plant species that are better adapted to high nutrient levels.

When most of us think of species extinction, we tend to think about the big, charismatic species that we feel an affinity towards—species like tigers, gorillas, and whales. But plants and insects form the backbone of biodiversity on the planet. To a certain extent, everything else, including us, relies on them. Humans evolved at a time of plenty on Earth. A stable climate, bountiful natural resources, high levels of life diversity, and vibrant ecosystem services all helped provide us with everything we needed to develop our modern society. By degrading these services and driving so many species to extinction, we put our own future in peril.

The good news is that if humans are causing the problem, we can still fix it. It won't be easy, but it's not impossible either. In Canada, many municipalities are banning the cosmetic use of insecticides and herbicides. That will help. As we learned from the Great Britain plant study, reducing air pollution and developing more sustainable agricultural practices will help too. Well-constructed legislation to protect endangered species can be an enormous asset as well. Such measures are a start, but we have along way to go. If butterflies and plants are indeed the real canaries in the coal mine, we don't have any time to lose.

CAN'T RELY ON CAPTIVE BREEDING TO SAVE SPECIES

One of the methods often proposed to save endangered species is to breed the threatened animals in captivity and then release their offspring back into the wild at a later time. The theory is that giving the animals a safe place to breed and rear their young makes it more likely that the species will be able to successfully survive in their natural habitats.

It seems simple enough. Of course, in nature things are never quite that simple. And recent research has found that breeding wild animals in captivity actually has consequences at a genetic level that make them less likely to be able to survive in the wild.

The problem is natural selection. As Charles Darwin described way back in his 1859 treatise *On the Origin of Species by Means of Natural Selection,* heritable genetic traits that are favorable to a species in a given environment become more common over successive generations. Unfavorable traits become less common. That's natural selection.

If you take an animal (or any living organism, for that matter) out of its natural habitat and introduce it to someplace new, natural selection takes over and traits that are favorable to the new location—in this case captivity—become more and more common in subsequent generations.

This poses a problem for animals in captivity because, try as we might, it's virtually impossible to replicate natural conditions in a zoo or recovery center. Space is one factor. Many animals require large amounts of space in which to roam, swim, or fly. Zoos can't be that large. In captive spaces the animals are also always safe, well fed, and made comfortable. Such "easy" living may make for physically healthy animals that are able to reproduce more often, but it doesn't necessarily make them genetically fitter.

In fact, quite the opposite. In a 2007 edition of the journal *Molecular Ecology,* Dr. Richard Frankham of Australia's Macquarie University explains how rare, and often detrimental, genetic variations that are suppressed in the wild can become common in captive creatures. These genetic adaptations to captivity have been documented in mammals, fish, plants, insects, and even bacteria. And while reproductive fitness often improves a great deal, Dr. Frankham points out that "Characteristics selected for under-captive conditions are overwhelmingly disadvantageous in the natural environment."

From turkeys to amphibians, fish, plants, and insects, it seems that the longer an organism stays in captivity, the less fit it becomes for the great outdoors. Fish from salmon hatcheries, for example, evolve smaller eggs that are less likely to survive in the wild. This helps to explain why reintroducing captive-bred species to the wild often has such a low success rate.

It doesn't help that while captive breeding is being conducted, the natural habitat of the captive species continues to be destroyed or otherwise made less hospitable to its former residents. In 2007, the British Columbia government announced that it is going to start a captive breeding program for the province's critically endangered spotted owl. Yet logging will be allowed to continue in the owl's habitat.

The loss of species from an ecosystem can also trigger changes that make the ecosystem itself less hospitable to the very species you are trying to recover. For example, certain large-seeded fruit trees rely on large primates to disperse their seeds, while the primates rely on the fruit for food. When primates are hunted to near extinction, the forest itself suffers and the fruit trees die off, so there's less food—making the reintroduction of large primates to the area even more difficult.

Saving a species from extinction may well require captive breeding. But it's really a Hail Mary pass to the future. Without habitat, a species is a mere caricature of what it is in the wild. The longer creatures are in captivity, the less likely they will be to ever survive on their own, even if we manage to stop destroying their homes. And that makes saving their homes in the first place that much more important.

BIG GAME CAN HELP PROTECT
ECOSYSTEMS FROM GLOBAL WARMING

Polar bears aren't exactly living large these days. Not only is their habitat shrinking because of global warming, but so are their genitals—thanks to industrial pollutants.

A 2006 paper published in an online edition of *Environmental Science & Technology* detailed the bears' most embarrassing plight. Researchers with the National Environmental Research Institute in Denmark looked at the genitals of one hundred polar bears from Greenland and found that the higher the levels of certain industrial pollutants in their systems, the smaller their genitals, and therefore the less likely they were to be able to successfully reproduce.

Polar bears ingest these toxins from the fat of seals and other marine mammals. Although areas of the Far North, like Greenland or Canada's Arctic, may seem pristine and far from polluted cities, long-lasting toxins are carried there by atmospheric and ocean currents. Once there, they accumulate in the food chain, eventually concentrating in the fat of marine mammals.

While dramatic stories about shrinking genitals may make headlines, far less attention is being focused on the role of top predators like the polar bear in the overall functioning of an ecosystem. That role is actually essential. If these large carnivores disappear, it can dampen the entire ecosystem's capacity to adapt to change—especially to big changes like global warming.

On the surface, losing a big predator might seem beneficial for other creatures in an ecosystem, which would otherwise be preyed upon by the giant carnivores. But the impact of big predators is much more complex. In nature, every creature has a role to play. The presence of large carnivores actually acts as a "top-down" influence that can benefit the entire food chain.

A 2006 paper published in the journal *Trends in Ecology & Evolution* provides an excellent example of this function. Researchers looked at the effect of wolves on the food chain in Yellowstone National Park. The gray wolf had been driven to extinction in the park in the 1920s but was reintroduced in 1995. Before their reintroduction, elk mortality in the winters had been dropping because winters had been getting shorter and less severe—likely because of global warming. This meant that less and less carrion was available for scavengers like coyotes, eagles, and ravens, which depend on carcasses for food.

However, once the wolves were established, they became the primary source of elk mortality throughout the year—increasing the availability of food for carrion-dependent scavengers. Computer models of food chain dynamics coupled with empirical evidence have enabled researchers to conclude that "the presence of wolves will enable scavengers to adapt to the effects of global warming over a larger timescale than if the wolves were absent."

In other words, top predators like polar bears or wolves can help entire ecosystems adjust to a changing climate. The paper concludes: "Their results clearly show that restoring top predators could be crucial for buffering the effects of global warming and also for reducing uncertainty in an increasingly unpredictable and warmer world."

Large carnivores are under increasing pressures from global warming, agricultural and urban expansion, unsustainable hunting, logging, and pollution. They also tend to reproduce slowly, and they need large spaces to survive. But if their presence can act as a buffer against change for entire ecosystems, then it makes practical sense to be working to conserve them and restore their abundance.

That isn't happening right now. Although the American government listed polar bears as a "threatened" species under the U.S. Endangered Species Act in 2008, the listing will have no effect on the rising American greenhouse gas emissions that are believed to be warming the planet and threatening polar bear habitat. In Canada, home to two-thirds of the world's polar bears, carnivores at the top of food chains have been routinely rejected for listing under Canada's Species at Risk Act, despite the recommendations of the scientific body responsible for Canada's threatened species. In 2008, Canadian polar bears were finally provided with the minimal status of "special concern," but this status provides little legal protection. As is usually the case with Canada's toothless endangered species legislation, politics trumps science. That has to change, because protecting large predators may prove to be an effective way to buffer entire food chains against the growing threat of global warming.

. . .

PARKS ONLY THE START OF CONSERVATION

People love parks. And why not? Even the smallest park can provide refuge from the concrete, steel, and plastic that otherwise dominates much of urban life. Sounds of wind blowing through trees, birds chirping, and squirrels chattering are a welcome respite from car horns and jackhammers. On a larger scale, parks that protect ecosystems from urban development, logging, and hunting can also protect the diversity of life that we all ultimately depend on for a healthy planet.

With polls telling us that public concern for the state of the planet is very high, some politicians are scrambling to announce new parks and wildlife areas. On the surface, who can argue with that? We certainly need more parks. But when I give talks about protecting nature and the services it provides, I tend to point out the necessity of changing human behaviors to staunch the loss of species and ecosystems. Sometimes this confuses the audience, and questions come up later. Reducing pollution? Slowing climate change? How would these measures help? Shouldn't we first focus on creating more parks to keep nature healthy?

This is a common perception about humanity's relationship with the natural world. It is also dead wrong. We cannot put a fence around nature and call it a day because we are as much a part of nature as any other creature. We depend on the services nature provides, like cleaning our water and maintaining a stable climate and fertile soils. For humanity to be healthy, the natural world has to be healthy, and the world can't be healthy if most of it is paved, with a few green spaces left "for the environment."

Still, it's a common misconception that environmental salvation can be had by simply putting dotted lines around tiny areas on a map where humans aren't allowed to run amok. The trouble is, nature doesn't work that way—she doesn't conform to our agenda. Nature is a complex, interconnected system made up of countless (and as yet many undescribed) parts. You can't isolate a small area, call it a park, and expect those boundaries to be honored by pollution or climate change. Nor can you expect various species to respect those artificial boundaries.

Parks are certainly important, but we have a long way to go before the world's parks could be considered truly effective at global species conservation. An article published in *Nature* reported that while protected areas now cover 11.5 percent of the planet's land surface (but less than 1 percent of the oceans!), these land areas are simply insufficient to protect biodiversity on a large scale.

As I wrote in an editorial for *Science* a few years ago, stemming the loss of life's diversity on the planet cannot be accomplished with a simple percentage target for protection—as politically expedient as that may be. Rather, effective protection—and I use the term "protection" to refer to any ecologically sound habitat management system, not simply the creation of parks—must occur and be enforced at all scales. In other words, in everything we do, we must try our best not to disrupt the natural systems around us because, ultimately, we are completely dependent on them. This is what sustainability is all about.

If we truly hope to ever live in balance with the natural world that sustains us, we really have to get over ourselves. Forces bigger than humanity are at work—forces that we mess with at our own peril. A healthy, vibrant future for humanity depends on the recognition that our best interests in the long term depend on maintaining healthy ecosystems. Moving towards this goal of ecological sustainability does not mean enduring hardships; it simply means having the foresight to act in our best interests now rather than trying to clean up our messes later.

Parks are a key element to conserving biodiversity and maintaining healthy ecosystems, but they are not the be-all-end-all—they are only the beginning. Trying to capture the Earth's life diversity in a series of randomly appointed parks and then attempting to keep those areas exactly the same forever will ultimately fail as a conservation measure. We already have places where such pieces of history are captured for posterity. They're called museums.

. . .

TWO LOST WORLDS GIVE US HOPE

In February 2006, two lost worlds hit the news. One was discovered halfway around the world, but the other is right here in North America.

The first was a never-before-examined patch of tropical rainforest deep in the heart of New Guinea. It's considered to be one of the most biologically diverse areas on Earth, and it shows how little we really know about life on this planet.

In 2005, an international team of scientists returned from the Foja Mountains of New Guinea after discovering forty extremely rare mammals (including the golden-mantled tree kangaroo, which was thought to have been hunted to near extinction), four new butterfly species, a new bird species, twenty new frog species, and many previously unknown plant species. Having never encountered humans, some of the creatures were so unafraid of people that researchers could simply pick them up off the ground.

That places such as this still exist is cause for hope. With well over six billion people on the planet and an insatiable appetite for resources, pristine places are becoming increasingly rare and species are disappearing at an alarming rate. Yet scientists have studied only a small percentage of life on Earth. Researchers estimate that there are literally millions of species out there that we have never examined and have no clue what they do in an ecosystem. As Oxford entomologist George McGavin points out, in a tropical rainforest, every second or third insect you pick up is probably unknown to science.

The other lost world in the news that week was also a remote and incredibly diverse rainforest—in Canada. British Columbia's unique north and central coast, known as the Great Bear Rainforest, contains creatures found nowhere else in the world. Most people know about the Kermode bears that live on this coast. They're a white version of the black bear, found only in this area. And their differences extend to more than just fur color: researchers are finding that they behave differently too.

Wolves of the Great Bear are also different—smaller, more agile, and specially adapted to forage for the bounty of sea life found along the shore. Then there are the salmon, which researchers have found are vital to the health of the forests and many land-dwelling creatures. Hundreds of unique runs of salmon find their way back to the Great Bear every year to spawn; their bodies provide nourishment to the wildlife, the trees, and the soil.

The Great Bear Rainforest made international news because the B.C. government, along with First Nations, environmental groups, and the forest industry, drafted a plan to protect a portion of it. It was good news for science and good news for the people who depend on the health of this ecosystem for their livelihoods.

Since the announcement, the groups involved have been working together to negotiate exactly how these changes will be implemented on the ground, particularly what kind of logging will take place in the parts of the Great Bear outside the protected areas. This is critical because unprotected areas make up more than 70 percent of the land base and contain the majority of salmon streams and much of the best wildlife habitat. By 2008, only some of the gaps in the agreement had been addressed, and many questions were still unanswered. In the meantime, logging in much of the area continued.

In 2009, many of the remaining gaps in the initial 2006 agreement are expected to be filled. I certainly hope that happens, for the sake of science and the future of the area. Scientists have only just begun to understand this magnificent region and all the life within it. The Great Bear Rainforest Agreement, if combined with truly sustainable logging practices outside the protected areas, could keep this ecosystem functioning, allow economic activities such as tourism and logging to coexist, and give scientists a chance to understand more about Canada's own lost world.

. . .

PREVENTING EXTINCTION BEFORE IT BEGINS

When most of us think about protecting species, we tend to envision ways of saving those creatures that are most endangered or whose populations are perilously low. This makes perfect sense—it's a kind of species triage, if you will. But it's also the most expensive and least effective way to prevent species from disappearing.

Far more effective would be to first identify which species are doing fine now but are likely to be in trouble in the near future. This would enable us to create more cost-effective conservation plans that will keep populations at healthy levels, rather than just on life support. But how do you identify those species on the edge?

A 2006 report published in the *Proceedings of the National Academy of Sciences* did just that. Researchers looked at available databases for information on nearly four thousand non-marine mammal species, then mapped the global geographic distribution of what they call "latent extinction risk." The results were surprising.

It turns out that the areas with the highest latent extinction risk are often not those that have a high number of endangered species—so they aren't often singled out for conservation. Most species conservation measures currently tend to focus on what have been called biodiversity "hotspots"— areas, usually in the tropics, that have very high levels of life diversity. However, this new research highlights a kind of "latent hotspot" where the potential for future loss of species is very high. And two of the largest areas in the world are in Canada and Alaska.

Why? Well, different species respond in different ways to human encroachment. Some can tolerate greater human disturbances than others. Those that don't tolerate human impacts such as road-building, hunting, oil drilling, and logging very well are more likely to suffer once people begin to move into their habitats. These creatures tend to be confined to relatively small geographic areas, have a large body mass, or be slow to reproduce.

It turns out that the boreal forest of northern Canada and Alaska, and the eastern forests of Ontario, Quebec, and the Maritimes, have lots of species with some of these characteristics—ungulates such as musk oxen and caribou, for example, or large carnivores such as polar bears or lynx. Some populations of these species have been fairly sheltered from human activity because of the vast size, challenging weather, and geography of these areas. However, as human activities encroach on their habitat, they are at risk of serious and sudden decline.

This precipitous drop is called "leapfrogging," and it means that a species previously considered robust suddenly passes other endangered animals and declines to the point of near extinction. It happened between 2000 and 2004 with the Guatemalan howler monkey when its critical forest habitat was destroyed.

In total, researchers for the *Proceedings of the National Academy of Sciences* paper identified twenty hotspots of latent extinction. Many of these areas are islands—such as New Guinea, Sumatra, and Borneo in Indonesia; the Bahamas; and the Melanesian Islands—where geography severely limits species' ability to withstand human pressure.

Canada and Alaska may have two of the largest areas of concern for latent extinction risk, but these jurisdictions are lucky in that they have small populations and large land bases. They have the resources and the time to develop conservation plans that take this research into account.

A latent hotspot analysis cannot help bring back the myriad species that are already endangered but, critically, it can help us prevent species from becoming endangered in the first place. It offers policy-makers new, potentially less costly, and more effective options to protect biodiversity. And that's an opportunity that we'd be foolish to ignore.

CROCODILE HUNTER MORE THAN JUST A SHOWMAN

Scientists sometimes call them "charismatic megafauna," but most people would just say they're cute and cuddly. Certain animals like bears, tigers, and the great apes have become poster children for the environment because, for many people, they symbolize the beauty and majesty of all nature.

Steve Irwin was not one of those people. Mr. Irwin, the famous Crocodile Hunter, was killed by a stingray in September 2006 while diving off the coast of Australia. He became famous not for showing the world the cutest and cuddliest of creatures, but for highlighting those that terrify us the most—crocodiles, snakes, spiders, and other creepy crawlies.

When Mr. Irwin died, I was in Australia on a book tour and was scheduled to meet up with him the following weekend. Sadly, that meeting will now never take place, and I will miss out on spending time with someone for whom I feel a great deal of kinship and respect.

Growing up in Canada, my passion and my playground was a swamp near my home. There, I waded through cattails to catch frogs, fish, spiders, snakes, and anything else I could get my hands on. I was utterly fascinated by these creatures and had a burning curiosity to find out what they did, how they lived, what they ate, and what ate them. I would not be surprised if Mr. Irwin had similar experiences as a child. Both of us seem to like things that others might call ugly or dirty. To us, they are all beautiful.

Certainly, I understand why people gravitate towards the most attractive, lovable creatures. It can even be beneficial and educational. Piquing people's interest in the environment with the world's most charismatic creatures may start them on the road to understanding and respect for all of nature. After all, *March of the Penguins* would never have become the international sensation that it did had it been called *Flight of the Turkey Vulture.*

But that's precisely what made Steve Irwin's role so important. True, he often went after the spectacular creatures himself—just not the pretty ones. At least, not pretty to most people. He went after the ones that were either unknown or vilified, hunted down, and despised by most of humanity. He's been criticized for doing this simply for the rush or to feed his ego, but in so doing he put the spotlight on creatures that would otherwise have been seen by the general populace only in our nightmares.

Every creature has a role to play in an ecosystem. Ugly, "dirty," or microscopic ones are often the most important. It has been said that all humans could disappear off the planet and the rest of nature would flourish and thrive, but if ants disappeared, the natural world would be thrown into chaos.

Humanity will not protect that which we fear or do not understand. Steve Irwin helped us understand those things that many people thought were a nuisance at best, a horror at worst. That made him a great educator and conservationist. At a time when interest in the basics of science, like taxonomy—the discovery and classification of living things—

is waning in favor of high-tech fields, it's a role that will be sorely missed.

Famed Harvard biologist E.O. Wilson coined the term "biophilia," meaning an innate love for, and kinship with, other biological creatures. Mr. Irwin had it in spades and he wanted to share it with the world. It was his enthusiasm for life on this planet—any life—that made him so remarkable. Steve Irwin may not have focused on the charismatic megafauna of the world, but the world clearly saw many of those same characteristics in him.

NEW SPECIES GETS FUNNY NAME

I've been called pond scum before, but never before has it been quite so accurate.

Researchers have discovered a new species of fly in the wet tropical rainforests of Costa Rica and Panama. They named the fly *Dixella suzukii*—after me. That's right, no great whale or fierce jungle cat for this fella. I've been immortalized as a fly. A pond-scum-sucking fly at that.

For many people, having such a fly named after them may not be such a big deal. In fact, some might actually find it a bit insulting. But for me, it is a great honor. Flies and I go way back. As a boy, I was an avid insect collector. And during the decades I spent as a working geneticist, fruit flies were my specialty. I spent countless hours in the lab experimenting with these remarkable little creatures. They are convenient to work with because they are so small and have such short life cycles. And genetically, fruit flies are surprisingly similar to human beings. It wouldn't be an exaggeration to say that I owe the bulk of my academic career to these insects.

But *Dixella suzukii* is no fruit fly. Like the fruit fly, it is a true fly of the order Diptera (along with 120,000 or so other species), but the fly that bears my name prefers to feed off of the scum that forms on the surface of swamps and slow-moving streams rather than fruit. In fact, as larvae and as adults, these flies are never far from a body of water.

Again, the similarity is uncanny. I, too, prefer to be close to the water. When I was a larva—I mean a child—you couldn't get me out of the swamp or the Thames River near my child-hood home in London, Ontario. True, I didn't eat pond scum, but that swamp and that river were certainly sources of nutrition from all the fish I caught there. Those small patches of nature cemented my love for the wild and for biology and ulti-mately propelled me down my career path as a geneticist.

So having a swamp-loving fly named after me couldn't be more apt. I owe this great honor to Sr. Luis Guillermo Chav-erri Sánchez of the Instituto Nacional de Biodiversidad in Costa Rica and Dr. Art Borkent of the Royal British Colum-bia Museum and the American Museum of Natural History. These two researchers found the fly, named it, and published their findings in the journal *Zootaxa.*

In the region of Central America where they found my *Dixella,* more than a dozen other new fly species were also found—yet only two species had been previously recorded in the area. This goes to show just how much we have to learn about the diversity of life on our planet—especially insects. Only about half of the world's fly species have been docu-mented so far. There are likely many more that we don't even know about.

In fact, *Dixella suzukii*'s home is actually one of most species-rich places on the planet. It's part of what research-ers call the Mesoamerica Hotspot. "Hotspots" of biodiversity are areas where an inordinate number of species live. Many of them are in tropical regions and face serious threats such as deforestation and global warming. The Mesoamerica Hotspot includes Costa Rica's vaunted cloud forest, which was once home to the golden toad—a spectacular species that went extinct in 1989. It is thought to be one of the first extinctions caused by global warming.

Dixella suzukii may just be a fly, but it is part of the great diversity of life on Earth. Each creature has a role to play in that diversity—roles that together make ecosystems healthier and more resilient to change. Maybe you can lose a species and the ecosystem can continue to function pretty much normally. Maybe you can lose two species. But at some point, the resiliency breaks down, and the loss can damage the entire system.

Sure, maybe having a bear or a shark or a raptor named after me would have been more dramatic. But insects are the workhorses of the natural world and are really just as important. So you can keep your elephants, your lions, apes, and other charismatic megafauna. I'll take my fly any day.

4

PUTTING MOTHER NATURE ON THE **PAYROLL**

Natural services and economics

YOU GET WHAT YOU pay for. It's one of those old, well-worn clichés that stays with us over the decades because of the simple fact that it's true. If you always look for the cheapest way to do something, regardless of other factors, quality will inevitably suffer.

Yet despite every businessperson, politician, and economist on the planet being able to recite this old saw by heart, that's not the way we've been treating Mother Nature. In fact, we've been stiffing her for a long, long time.

It turns out that in the mad rush to create a global economy, we forgot one thing—the globe. This spinning orb we live on happens to provide the base materials for everything that our ingenious human minds can create. Even more important, the Earth's natural systems create the necessary conditions for life on the planet to exist at all. They keep the air clean, the water fresh, and the soils fertile.

You'd think that if we valued these things, we'd have some way of accounting for them. But we don't. Mother Nature has been incredibly generous, offering these services for free. For our part, we've accepted them gladly and without question. In fact, we predicated our entire economy around the notion that these services have no quantifiable value and will continue to be provided indefinitely.

As this series of essays explores, we don't have that luxury anymore. Our knowledge of Earth's systems and services may be imperfect, but we know that many of our actions are damaging them. The planet can provide only so many resources and absorb so much of our waste. After that, things start to break down. And breaking down they are.

Nowhere is this more obvious than with global warming. Exhaust gases such as carbon dioxide from the tailpipes of our cars, the chimneys of our homes, and the smokestacks of our industries are building up in the atmosphere, enhancing the Earth's natural greenhouse effect and trapping more heat near the planet's surface, like a blanket. But this extra heat and energy doesn't make things all warm and cozy. Because the atmosphere is a system, it's more like throwing random weights on a precariously balanced load, unbalancing the system, and disrupting a climate that's otherwise been relatively stable.

Dealing with this problem will be expensive. Very expensive. And the longer we wait to fix it, the more expensive it will be to deal with later on. That's why it's so important to come up with a way of valuing our atmosphere—such as a carbon tax—immediately. While it would be nice to be able to simply say that nature's services, like providing fresh water and a stable climate, are priceless (they do, after all, keep us alive), our society is ruled by the almighty dollar. The sad fact is, unless we are able to put a price on these services, we will continue to squander them.

Does this mean that we must assign dollar values to all of our experiences in nature? How much is inspiration worth? What is the value of knowing there are wild places left in the world? Of knowing that our closest cousins, the great apes, still inhabit our Earth? Of being able to honestly say that we have been responsible stewards of the only planet known to harbor life in the universe?

A free-market economist might say that these things are worth exactly what people are willing to pay for them. But the tragedy of the commons is that most people don't even know what natural services are, let alone how their actions are damaging them. And even if people do, they want to be quite sure that their neighbors are also paying their fair share for services that benefit us all. This is why government policy is critical— to create a level playing field where those who pollute and damage nature's services pay more than those who do not.

As fair as that may seem, powerful people and industries want to maintain the status quo, arguing that taxing them will hurt their bottom lines and weaken the economy. They're right, of course, unless you factor in the value of our natural services—which is the whole point. Welcome to Earth, the only planet we have. The services she provides are valuable. And someone, somewhere, has to pick up the check.

. . .

PUTTING A PRICE TAG ON NATURE

Many people believe that nature's value cannot be put into dollars and cents. That is, they value the natural world for its own sake, regardless of what services or benefits it provides for humans. Yet this notion is fundamentally at odds with the economic system we've created.

We live in a world increasingly dominated by a global economy, where it is assumed that everything of value has a price tag attached. If something can't be quantified and sold, it is considered worthless. The CEO of a forest company once told me, "A tree has no value until it's cut down. Then it adds value to the economy."

So how do we reconcile our economy with ecology? The Earth provides us with essential natural services like air and water purification and climate stability, but these aren't part of our economy because we've always assumed such things are free.

But natural services are only free when the ecosystems that maintain them are healthy. Today, with our growing population and increasing demands on ecosystems, we're degrading them more and more. Unfortunately, remedial activities and products like air filters, bottled water, eyedrops, and other things we need to combat degraded services all add to the GDP, which economists equate to progress. Something is terribly wrong with our economic system when poor environmental health and reduced quality of life are actually good for the economy!

But what if we did put a price tag on things like clean air and water? If we assigned a monetary value to natural systems and functions, would we be more inclined to conserve them? This idea was pretty radical when it was first seriously proposed in 2000 by an international group of ecologists writing in *Science*.

The group argued that humanity will continue to degrade natural systems until we realize that the costs to repair or replace them are enormous. So we must find a way to put a dollar value on all ecosystem assets, including natural resources such as fish or timber, life-support processes such as water purification and pollination, and life-enriching conditions such as beauty and recreation.

Most of these assets, with the exception of natural resources, we already exploit but do not trade in the marketplace because they are difficult to price. But this is changing.

In 1997, for example, New York City officials decided to begin buying land around watersheds and let the forest and soil organisms filter water instead of building a massive new filtration plant. It wasn't that city planners were closet environmentalists trying to preserve nature. The economics just made sense. Protecting a service that nature provided for free was far cheaper than engineering a Band-Aid solution to clean up the water afterwards. Until recently, this kind of potential to use natural services rather than technology to solve problems has been largely overlooked, even though natural approaches may provide greater benefits to communities such as lower costs, reduced flooding and soil erosion, and aesthetic benefits.

In Canada and the U.S., forests are primarily valued for the timber they provide. But this leads to conflicts. For instance, a report from the Canadian Department of Fisheries and Oceans found that logging roads in the province of British Columbia continue to devastate fish-bearing streams, even though legislation is supposed to protect them. In fact, these forests provide many services, including habitat for plants and animals, recreation, and others that, if assigned a monetary value, could completely change the way we use them. In Australia, the New South Wales Department of Primary Industries promotes the future of forests as being tied to ecosystem services, with timber considered simply one of the many products and services that intact forests can provide to human beings.

As just one species out of perhaps fifteen million, the notion of assigning value to everything on Earth solely for its utility to humans may seem like an act of incredible hubris. But the harsh reality of today's world is that money talks and economies are a central preoccupation. At the very least, assigning monetary value to ecosystem services may force us to take a hard look at all that nature provides. Maybe then we'll stop taking it for granted.

* * *

CONSERVING NATURE IS LIKE MONEY IN THE BANK

Ask folks what they value about nature and most would probably be quick to mention aesthetic and spiritual properties like beauty, serenity, and peace. We hold these values dear to our hearts because they resonate with strong emotional ties. But there are other, more pragmatic, reasons to value nature—reasons even a hard-headed economist can't deny.

We've lost touch with the fact that everything we have depends on nature. Without the rest of nature propping us up, we could not survive—a fact so obvious that it seems silly to point it out. The problem is, we don't behave as though this were obvious. We behave as though the economy is completely separate from the world in which we live. Industrialized society is geared entirely towards output—how many Wiis, suvs, and cans of Pepsi we can create, sell, and consume. What aren't factored into the equation: the natural services needed to support this output. Why? Because nature's services are considered free.

And in a standard economic sense, they *are* free. Nature is the source of clean air, water, and fertile soil with no strings attached. However, with over six billion of us now shuffling up to nature's buffet, the "all you can eat" sign will have to come down soon or those at the back of the line—the next generation—will be left with nothing but Jell-O salad.

Efforts to quantify the value of nature's services have been met with suspicion by some economists. In 1997, an international team of researchers headed by ecological economist Robert Costanza came up with an estimated annual average of about US$33 trillion—roughly twice the size of the annual GDPs of all the countries in the world, and virtually none of it accounted for in the marketplace. Their paper, published in *Nature,* was the first of its kind. But it only looked at the overall monetary value of natural services. It didn't answer the question of what effect human activities are having on nature's "net worth."

So, in 2002, a group of researchers attempted to tackle this question. The results were shocking. Their analysis, presented in *Science,* showed that simply in terms of dollar value, conserving natural areas is actually *one hundred times* more profitable than exploiting them.

The researchers looked at five real examples—logging in Malaysia, small-scale agriculture in Cameroon, mangrove swamp conversion for shrimp farming in Thailand, drainage of marshlands for agriculture in Canada, and the destruction of coral reefs by dynamite fishing in the Philippines. In each case, the economic value of the conversion activity—such as the sales of the end product and jobs created—was far less than the value of the services provided by intact natural habitats nearby (things like sustainable, low-impact logging, flood protection, sustainable hunting and fishing, and provision of clean water). In total, the researchers estimated that worldwide loss of natural habitat costs humanity some US$250 billion every year. And because the conversion is permanent, those losses continue every year into the future, in addition to the next year's losses.

Such outrageous costs immediately raise the question: if these practices are so uneconomical, why hasn't someone stopped them? The answer is that the savings associated with conserving nature are spread throughout society, whereas the profits earned from exploiting natural resources are immediate and benefit a narrow group of individuals. Many current government subsidies and tax incentives also support such practices. In fact, researchers estimate that these subsidies add up to between US$950 billion and US$1,950 billion every year.

Our global economic system has been constructed under the premise that natural services are free. We can't afford that luxury anymore. We have created a deeply flawed system, but we can still change it. With new knowledge of the extent to which we are mortgaging humanity's future by subsidizing narrow economic interests, conventional views on economic development must be reconsidered and reconstructed to make ecological conservation a priority. We must put the economy in synch with the natural world that made it possible.

CONSUMER PRODUCTS START AND END WITH NATURE

Christmas season means advertising season. Local mailboxes and newspapers laden with flyers. Children pleading for the latest toys they've seen on television. Awash in consumer products and commercials, it's easy to forget where all this stuff comes from and where it goes.

We put up with advertising saturation because we don't have much choice. Between television and radio commercials, print advertising, pop-up internet ads, and billboards, we can't get away from it. Some companies will help pay for a new car if you're willing to drive one festooned with advertisements. For five thousand dollars a Texan even agreed to have an advertisement permanently tattooed on the back of his head. Scary stuff.

But every once in a while I still see an advertisement so fundamentally stupid it takes my breath away. Such is the case with a television ad for a dishcloth that has been impregnated with soap. The big kicker? It's disposable. Use it once and throw it away. Why not? As the actor in the commercial says, "There's plenty more where that came from!"

Funny, I hadn't realized that plain old reusable dishcloths were such a terrible inconvenience. Good thing a nice company was willing to point out my hardship. And look, they have a product to solve the problem. How thoughtful!

Now, this product by itself is not going to cause environmental disaster. In fact, disposable products like this are pretty minor in the big scheme of things. Humanity's largest impact on the environment comes from things like extracting raw resources such as oil, driving our cars, heating our homes, and growing and processing our food. Disposable cloths barely register on the radar screen of environmental problems.

It's the attitude that is the real problem: "there's plenty more where that came from." This, in a nutshell, is the reason why humanity has managed to get itself into so much trouble. We keep assuming there's plenty more of everything—plenty more fish in the sea, plenty more forests, plenty more wild spaces, plenty more natural resources, plenty more room in the oceans and in the atmosphere for our wastes.

In reality, our planet is actually very small and interconnected. Our atmosphere, for example, isn't much more than six miles thick. So it shouldn't really surprise us that spewing air pollution and heat-trapping gases from our cars and factories causes smog and global warming.

These problems are costing us a fortune. According to the Ontario Medical Association, smog costs taxpayers in Ontario alone more than c$1 billion every year through increased health care costs and lost workdays. Global warming expenses are just beginning, but record extreme weather events around the world in 2005 caused more than US$200 billion in damages—something the World Meteorological Organization says will continue to increase with global warming. For years now, insurance companies have been telling us that their payouts for weather-related damage are skyrocketing.

If we want to change this trend and have a clean, healthy, sustainable future, we have to consider the real costs of all our activities and products. We have to build external costs into the price of everything from gasoline to disposable dishcloths. We have to encourage clean practices and efficiency and discourage waste.

Right now, we are very wasteful with our use of natural resources—including energy—largely because society as a whole, rather than an individual or a company, has to pay for the resulting pollution. The company touting disposable dishcloths, for example, does not have to pay for increased air and water pollution, diminishing resources, landfill waste, rising health care costs, and global warming. We all do. Consider that when you see a commercial for a blatantly wasteful product. Plenty more where that came from? Think again.

*   *   *

ECONOMY NEEDS A BETTER GOAL THAN "MORE"

It's one of those questions that drifts in the shadows of our modern world, just waiting to be asked: "How much is enough?" Yet few people ask.

Under our current economic system, you can never have enough and you can never have too much. In fact, our entire economy is predicated on continued, endless growth. Yet we live in a finite world, with finite resources and a limited amount of space to dump our wastes.

Bit of a problem there.

In fact, right now, the standard measurement of human well-being is Gross Domestic Product (GDP), that is, the monetary value of all goods and services produced by a country. If the country's GDP is high, then well-being is considered high. If the GDP is stagnant or, god forbid, declining, then—regardless of other indicators—politicians go into crisis mode to "get the economy moving again."

Of course, one immediately wants to ask, "Where, exactly, is it going?" To which the answer is always "Up!"

There is a good rationale for all this in that economic growth is tied to jobs and income, which are indeed to a certain extent tied to well-being. But the GDP also includes things like cleaning up oil spills, clearing car accidents, and treating asthma attacks brought on by smog. And it includes things like strengthening process efficiencies to "improve the bottom line"—which actually means laying off workers so shareholders make more money. Is that really good for well-being?

Few people today seem to be asking the fundamental questions: how much is enough? And what is an economy for? In the post-WWII United States, consumption was seen as a way to make sure a wartime economy could remain in high gear in peacetime too. According to the chair of President Eisenhower's Council of Economic Advisors: "The American economy's ultimate purpose is to produce more consumer goods."

So, this thing that pretty much rules the world and dominates politics at all levels, this thing that citizens are expected to submit to virtually without question and "help prop up" or "buckle down for" or whatever we are being told to do at any given time, the ultimate purpose of this thing that so controls all of our lives is to . . . create more stuff?

This seems beyond narrow-minded—it's dangerous. It's putting us on a fool's path to disaster by distracting us from very real environmental and social problems, allowing us to shrug them off with a simple: "What can you do? It's the economy!"

But we can do something. Our economy is a social construct and right now it's not working for us; we're working for it. We have no goal and without one, we will never be satisfied and never know when enough is enough.

According to ecological economist Robert Costanza, if you make the goal of an economy sustainable human well-being rather than growth, it allows you to consider a comprehensive suite of things that should be brought into economic decision-making—things like the value of natural services, education, and fulfilling employment.

Costanza groups these into four types of capital: built capital, human capital, social capital, and natural capital. He contends that all of them are important elements of an economy and are necessary to examine if we are to ever get away from the single-mindedness of relentless growth.

We have a population expected to reach nine billion by 2050, a limited supply of natural resources, looming environmental concerns, and an economy whose sole purpose is to produce more and more stuff. This is a problem. It isn't working, and it's time to find something else for our economy to do.

.   .   .

MEETING REGULATIONS COSTS LESS THAN EXPECTED
In 2001, when U.S. president George W. Bush decided to renege on his country's promise to reduce greenhouse gas emissions under the Kyoto Protocol, he used economic cost as the overriding concern, saying that trying to meet the agreement would put an undue burden on the American people. Five years later, when Stephen Harper became prime minister

of Canada, he too ditched the Kyoto Protocol (after it had been ratified by Canada's parliament), citing some industries' high estimates of meeting the targets as his rationale. It would just cost too much.

Or would it? Time and again, when it comes to meeting environmental regulations, the estimated costs have proven to be far higher than the actual costs, and the economic benefits downplayed. In fact, the history of environmental regulations tells us that the true costs of protecting the environment are never as high as industries claim they will be. From asbestos to benzene and from CFCs to sulfur dioxide, industry claims of economic disaster have not come true.

A study a few years back by the Economic Policy Institute in Washington, D.C., found that in almost every single case they looked at, the costs of complying with environmental regulations were far lower than industry—and even governments—claimed they would be. For example, electric utilities in the U.S. claimed that it would cost US$4 billion to $5 billion per year to meet the requirements of the 1990 Clean Air Act. But by 1996, utilities were actually saving $150 million per year.

CFCs are another example. When a phase-out of these substances that damage the ozone layer was announced, many industries claimed that alternatives did not exist or were too expensive. In 1993, car manufacturers said the CFC regulation would increase the price of all new cars by up to US$1,200. Just four years later, the industry admitted that costs were already down to as little as $40.

In perhaps the best-known Canadian case, the giant metal company Inco vehemently opposed reducing emissions from its smokestacks but, once forced to do so by government regulations, discovered money could be made from the material captured in scrubbers. It now boasts of its environmental awards and civic leadership.

Complying with environmental regulations is almost always cheaper than opponents claim for several reasons. First, much of the touted costs are for capital equipment that is usually much more efficient and cleaner than old, dirty equipment. These costs more appropriately should be considered capital investments, which end up reducing overall operating costs. Second, technologies change and improve, and once adopted on a mass basis, these technologies benefit from economies of scale that result in lower costs. Finally, in complying with regulations, industries are forced to rethink standard business practices that may have been wasteful or unproductive.

Had the United States and Canada embraced Kyoto, it would likely have proven to be just another example of a situation where the costs of meeting a target were blown out of proportion. Had North America's engineers and scientists turned their efforts to reducing greenhouse gas emissions, they may well have even discovered ways to make it profitable. As retired Canadian senator and forward-thinking oil industry businessman Nick Taylor said in an interview, "It is so transparently obvious that Kyoto is going to make [oil-rich] Alberta money that Alberta is bound to come around."

Unfortunately, that didn't happen. The Alberta government actually used public money to campaign against Kyoto, despite numerous studies indicating many economic benefits and opportunities resulting from the treaty. Prime Minister Harper, an Albertan himself who once called the Kyoto Protocol a "socialist scheme," obliged, dismissing the accord and watering down Canada's commitment to reducing greenhouse gases.

We'll never know what implementing Kyoto would have really cost North Americans, or the benefits it could have provided to our health, our economies, and our climate. Traditionally, we have viewed air, water, and soil as limitless resources for industrial use with little regard to long-term effects on ecosystems, health, or aesthetics. Indeed, economists generally consider air, water, and soil as "externalities," factors outside the economy. In recent decades, we've become aware of how much the massive use of fossil fuels is altering the atmosphere, affecting both climate through greenhouse gas accumulation and human health through air pollution. We cannot afford to ignore these problems anymore. Fortunately, it's still not too late. We have history on our side. We can still change, and it won't cost us the Earth.

·   ·   ·

ENVIRONMENTAL INCOME CAN HELP REDUCE POVERTY

A common complaint I hear again and again is that environmental groups ignore the economy. "You can't take care of the environment unless you have a strong economy," is the standard refrain.

This line is often used as a justification to pursue just about any development, regardless of its environmental impact. And it's often used as a club against those who seek to protect natural areas in the developing world, as though industries are simply seeking to better the lives of the poor in these areas while environmental groups want to hold them back.

Of course, nothing could be further from the truth. Alleviating poverty is an important environmental goal too. I have often said that if my family were starving and I saw an endangered plant or animal we could eat, I'd have to kill it and bring it back to my family for food.

Still, the argument is tempting in its simplicity. How else can we improve the lot of the poor if not by merely exploiting the natural resources where they live? A 2005 report, *The Wealth of the Poor*, backed by the United Nations and the World Bank, and a 2008 follow-up, *Roots of Resilience: Growing the Wealth of the Poor*, help explain how. They argue that linking aid to environmental protection is one of the most effective ways to reduce poverty.

According to the reports, economic-development programs have largely ignored the three-quarters of the world's poor who live in rural populations in developing countries. These people more than any others rely most directly on natural resources and services, yet they face increasing pressures on these services from outside forces. Fish stocks, for example, are dwindling in developing countries as industrial fleets mine the seas to feed wealthier people in the developed world. This economic activity might increase the developing nation's GDP in the short term, but it actually reduces the capacity of local people to earn a living over the long term.

In other words, many of the current models of economic development are ill suited to rural areas because they fail to take into account the connection between the people and the planet. But more than that, *The Wealth of the Poor* argues that rather than just being basic survival mechanisms, natural resources and services can actually be wealth-creating assets if they are effectively managed. This "environmental income" can act as a stepping stone to economically empower the rural poor.

Central to tapping into that wealth is the need to bring local resources under local control. Examples have shown that such measures can succeed where international industrialization schemes have failed. For example, communities in Fiji increased the abundance of fish in their waters when they rigorously restricted fishing to certain areas. And land reforms based on local community cooperation in northern Tanzania have led to reforestation in degraded areas that now provide food and fuel for local people.

Unfortunately, such examples are few because the vast majority of poverty-reduction schemes are not grounded in local people and their environment. Natural services are often ignored as though they have no value, and the poor suffer the most when these services are degraded. If we want to improve the lot of the rural poor, that has to change.

It's true that environmental organizations have not paid enough attention to poverty. But it's much more glaring that economic development plans have paid so little attention to the environment. For too long there has been a disconnect between science and economics. The fact is, we all depend on natural services and resources for our survival and for the wealth of goods we currently enjoy. And if we don't take care of those natural systems, we will all end up much poorer.

STERN WARNING CHANGES THE CLIMATE DEBATE

Sir Nicholas Stern's 2006 report on the economics of global warming finally changed the nature of the debate. Instead of being pigeonholed as an environmental problem, global warming can finally be seen for what it really is—an economic one.

The *Stern Review on the Economics of Climate Change* was huge. Literally, at seven hundred pages, but also for the shock waves it sent around the world. Mr. Stern is no lightweight. He's a former chief economist with the World Bank. At his disposal was a team of twenty other researchers and academics. And the report they presented, while certainly not the final word on the topic, has finally assigned a dollar figure to the costs of global warming.

That figure is astounding: us$7 trillion—or between 5 and 20 percent of the global economy—wiped out by the beginning of the next century because of problems brought about by a warmer planet. Global warming, the report says, could cost more than the two World Wars combined and lead to a worldwide depression.

Predictably, some people pounced on the report, saying it was alarmist and inaccurate. Of course, the reality is that no one knows how accurate it is. Creating a report of this kind naturally requires making certain assumptions and even some guesses about future trends. But it represents a very good estimate that can now be refined over time.

What we mustn't do, however, is get so bogged down in fighting over the details that we fail to see the most important message in the report—that we can't afford not to take serious action. For years, many politicians and industry lobbyists have painted global warming as an environmental problem—as though the solution could be found through traditional environmental management methods, such as creating a new park, helping an endangered species recover, or planting trees. Yes, we need to help the environment, they said, but we have to be careful not to do anything that could slow economic growth. In fact, some went to great lengths to insist that tackling global warming would mean "shutting down" the economy. Sure, they lamented, we could do something about the problem, so long as you don't mind living in caves and eating dirt.

Basically, global warming was painted as an either-or environmental problem. Either you had a robust economy and accepted a hotter planet that might not have as many pretty birds or plants, or you had no economy and lived like the Flintstones with lots of fuzzy animals and spotted toads. According to the standard argument, you couldn't have both.

Mr. Stern's report showed that this is a lie for two reasons. First, the economic costs, not just the environmental costs, of inaction are actually much higher than the costs of adequately dealing with the problem now. Mr. Stern originally estimated that measures to stabilize greenhouse gas concentrations at levels to prevent dangerous global warming would cost just I percent of the global GDP. By 2008, citing new evidence that our climate is changing even more rapidly than anticipated, he had doubled his cost estimate to 2 percent of the global GDP. Still, this means that we most certainly can afford to take significant action, and it highlights the urgent need to get started now, rather than waiting.

Second, the report showed that tackling global warming is not about saving the whales or some such thing; it's about not being stupid. It's about having the capacity to recognize that the health of the world we live in and the health of our people and our economies are intimately connected. It's about recognizing that, although we rarely think about it, the services provided by nature are worth a great deal (of money, if you like to think of it that way).

We can't stop global warming in its tracks, but we can avoid the worst of it. Fighting the problem certainly has a price, but it's manageable. These were the lessons of the *Stern Review*. Reasons enough to put aside the rhetoric, stop the posturing, and not be stupid.

* * *

BANKING ON OUR NATURAL CAPITAL

Anyone who regularly reads science journals knows there's no shortage of research about how human activities are affecting our ecosystems. But translating that research into action to conserve those systems is another matter.

Sometimes it seems dire predictions about the future of the planet's ecosystems come out almost every day. Then the headlines fade (if there even were headlines), and so does the sense of urgency. As a 2005 editorial in *Nature* points out, "In too many cases, however, that leaves scientists positioned only to track the loss of these systems. So far, researchers have been less effective at achieving the level of impact on policy decisions needed to implement actual conservation measures."

Standard appeals for conservation—such as the need to conserve species for future generations, protect especially charismatic species, or preserve the aesthetic beauty of certain areas—arguably aren't doing very well. By most measurements, society is failing to conserve the diversity of life on Earth, as well as the natural systems that provide many important services to humanity—things like helping stabilize the climate, clean our air and water, and keep our soils fertile.

If standard appeals aren't working, how do we best appeal to policy-makers to make the changes necessary to achieve sustainability? Well, one increasingly popular way is to stress the economic value of ecosystems. That may seem crass to some. After all, how can you put a dollar value on some services that are irreplaceable? How can you put a price tag on the planet's ecosystems that support us when it's the only planet we have?

Those are valid questions, but if policy-makers only understand dollars and cents and natural services are valued at zero, then nature will continually be overused and damaged by those who make the economic decisions. Natural services are economically valuable, and when we reduce their capacity to function, we lose out—in terms of both dollars and quality of life. Research shows, for example, that the terrible loss of life and property resulting from the December 2004 tsunami could have been lessened if mangrove forests along the Sri Lankan coast had been protected.

North American countries, too, rarely put a value on nature or the services it provides. Smog costs in Ontario alone are at least C$1 billion every year in health care expenses and lost workdays—and that number is increasing. These are actual hard costs to society. Yet Canada and the U.S. don't internalize the costs to polluters. Instead, they let polluters off without paying the bill, and society as a whole has to pick up the tab.

Some even suggest that taxes on gasoline should be reduced, further removing responsibility from the most polluting industries and individuals. This makes no sense from an environmental or an economic perspective. As the respected magazine *The Economist* points out, "Petrol taxes are there to capture and charge motorists and others for the externalities they create, such as pollution and congestion. . . To cut fuel taxes when oil prices rise is bad economics as well as bad politics."

Once you factor in the external costs to society of degrading natural systems, it's often a better economic investment to protect some ecosystems rather than exploit them in an intensive industrial fashion. For example, Canada's huge boreal forest, which stretches across the North, is one of the world's most important carbon sinks. It's incredibly valuable in helping to reduce global warming. If we attach a dollar value to this service we can see why it makes sense to protect large areas of this forest from industrial development and logging.

Trying to conserve little bits of land here or there and trying to protect the variety of life on Earth species by species just isn't working. The results of the first four-year study by the United Nations–sponsored Millennium Ecosystem Assessment told us that 60 percent of the planet's ecosystem services are currently being degraded by human activities. Focusing on the economic value of these services may well be the only hope we have of protecting them.

.   .   .

GOING INTO DEBT A RISKY PROPOSITION

Most of us are all too aware of what it's like to live in financial debt, but what about ecological debt?

On September 23, 2008, as financial markets around the world crashed, another important indicator of our debt load went by virtually unnoticed. According to the Global Footprint Network, that's when humanity went into ecological debt for the year. It's when demand for resources and the production of waste outpaced the planet's capacity to produce new resources and absorb those wastes. In other words, we ceased to live off the ecological services provided by the planet and started consuming the ecosystems themselves.

The date is merely symbolic, as in reality human consumption of resources and production of waste is highly varied across the planet. In some areas, we're already going into debt at the stroke of midnight on New Year's Day. Other areas, however, are far less exploited, and we may never reach those particular ecosystems' ecological limits during the year.

So ecological debt is more of a global average, based on the "ecological footprint" concept conceived by Bill Rees and Mathis Wackernagel at the University of British Columbia. The footprint concept is easy to visualize and helps us understand and compare our rate of resource consumption. North Americans' ecological footprint, for example, is huge. If everyone on Earth consumed as many resources and generated as much waste as we do, we'd need the equivalent of nearly five more planets!

Ecological debt is similar, as it also helps us understand how human activities are affecting the planet and the services it provides to us. The concept requires us to look at these services as if they are sorts of paychecks. If we live off our income, we're doing fine—that's sustainability. But when we start living beyond our means, just as we would with our finances, we go into debt and we may end up in trouble. Global warming is one example of that kind of trouble.

The Global Footprint Network calculates that the first ecological debt day occurred in 1987, on December 19. But every year since, it has been getting earlier and earlier as our rate of consumption has increased. By the mid-1990s, it had moved up to November. In 2006, it was October 9. And the march continues. What happens if we break the bank? Well, it certainly would be problematic for our species, as the planet could simply no longer provide all the services we need and absorb all the wastes we create. Our population would then have to shrink down to a level that was sustainable with whatever functioning ecosystems we had left. We could see an ecological crash akin to what happened to the financial markets in 2008.

Although we often talk as though we should reduce our impact on the planet to protect nature, it's actually much more about protecting ourselves. As pointed out in the excellent book *The World Without Us* by Alan Weisman, nature would get along quite well without us. If humans were to go ecologically bankrupt and die off as a species, nature would no doubt spring back.

Human beings may have permanently altered some ecosystems, but life on Earth is remarkably tenacious. Without people around, wooden structures in our cities would start to decay almost immediately, and plants pushing their way into cracks would gradually overtake concrete, turning sprawling suburbs into forests and prairies once again. Overfished seas would rebound with life. Many species currently on the brink would flourish. Our oceans would gradually absorb the carbon dioxide we've pumped into the atmosphere. Even nuclear waste would gradually decay. As Weisman points out, alien visitors to Earth a hundred thousand years after our demise would see no obvious signs of what we once were.

It's sad to think that all we have created on Earth could, in evolutionary terms, disappear in the blink of an eye. Life would go on, but the remarkable story of a unique bipedal species would come to an end. A humbling thought—and a compelling reason to stay out of debt.

TIME TO PUT A PRICE ON POLLUTION

Mention the concept of a new tax to politicians and most will run screaming out of the room to go vacuum their cars or mow their lawns—anything to avoid talking about an issue that they think could lose votes, no matter how sensible or reasonable the concept may be. But that's going to have to change soon, because we need to have a serious and open discussion about initiating a mechanism for pricing pollution—specifically carbon.

By now everyone's aware of the mounting challenges we face from global warming. The science, while still ongoing, is clear: the heat-trapping gases such as carbon dioxide that we're pumping into our atmosphere from our homes, cars, and industries are warming the planet and disrupting the climate. If left unchecked, the consequences will be severe—to our environment and our economy.

So it's in everyone's best interest to start curbing our carbon output. There are many ways to do this, but most experts agree that market-based solutions can play a critical role. Two such solutions are a cap-and-trade system and a carbon tax. Under a cap-and-trade system, governments put a limit on the amount of carbon that can be released into the atmosphere. Industries have to stay within their limits. Innovators who go below their limits can sell their leftover emissions as credits to those who go over the set amount.

Under a carbon tax, the more you pollute, the more you pay. Such a tax could be applied to all products or activities that have a substantial carbon footprint—producing and burning gasoline, coal, and other fossil fuels, for example. This would encourage industries to become more efficient and reduce costs while encouraging consumers to save money by being more environmentally friendly.

A 2008 report by the Canadian government–commissioned National Round Table on the Environment and the Economy found that, regardless of which mechanism we choose, the longer we wait to put a price on carbon, the more costly it will be. The report said that because businesses and investors make long-term decisions about capital costs like buildings, technologies, and equipment, they need a clear idea of where the government is heading: "In essence, inadequate and delayed communication by the government of a [greenhouse gas] 'price' could lead to substantial long-term economic costs."

Politicians have a knee-jerk reaction to taxation, as do many North Americans. However, I don't think North Americans feel taxes are necessarily bad so much as they think wasting tax dollars is bad and unfair taxation is bad. By its very nature, a carbon tax should be reasonably fair because it directly taxes the product that causes the harm and expense to society as a whole. The more you pollute, the more you pay. That seems pretty fair.

But North Americans would also revolt if they felt their tax money was being wasted. That's why it would be essential to dedicate the money gained from a carbon tax to developing and promoting more sustainable alternatives. Proceeds from a carbon tax could be put towards providing better public transit, for example, thus improving the service or reducing the cost of a more sustainable transportation option. For electricity production, proceeds from a carbon tax on say, coal, could go towards cleaner, renewable energy sources like wind.

Still, while the concept of a carbon tax makes perfect sense, and leading economists around the world support it, much of the North American public has yet to be convinced. In the province of British Columbia, Canada, the government has recently instituted a carbon tax. It's well-thought-out and forward-thinking. But it is widely unpopular among the electorate, who see it as another "tax grab." Similarly, in 2008, the leader of Canada's Liberal Party, Stéphane Dion, brought forward the idea of a carbon tax as a way to "make polluters pay" while reducing income tax for average Canadians. Again, the concept made sense, but Canadians didn't trust it. Many pundits say that Mr. Dion may have lost the 2008 Canadian federal election in large part because of this issue.

Global warming has really changed the environmental discussion in North America and in much of the world. Suddenly, people are much more aware of our environmental challenges and eager to get moving on sustainable alternatives. But while many people are aware of these issues and eager for change, they remain suspicious and mistrustful of their governments. That has to change. We need government involvement to get our countries pointed in the right direction. Governments play a key role in this movement. It's up to them to get the public on board to understand how important carbon pricing is for our future. If our federal governments aren't already seriously looking into a carbon-pricing mechanism, they should be.

5

HOT HOT **HEAT**

Global warming and climate change

FOR US PUNY HUMANS, few things could possibly seem as vast as our atmosphere. Like space itself, our skies seem limitless and our own actions within it inconsequential. Coming to grips with the notion that our behaviors have actually changed the composition of the atmosphere and influenced weather patterns and the climate—things that have traditionally been considered "acts of God"—well, that's a tall order indeed.

So the fact that it's taken as long as it has to bring the public to some level of understanding on this issue should not perhaps come as a complete surprise. But it's certainly been a struggle. Over the past decade, I have written about hundreds of topics, ranging from butterflies to politics and television advertising. But there is nothing that I have written about more than climate change. Thousands of words, tens of thousands: a measure of both the issue's importance and our inaction in doing something about it.

Part of the problem has been the inherent complexity of our climate. While most of us regard the air around us as a pretty simple thing, it's really quite the opposite. In fact, the air, the oceans, plants, and trees, and even our soils are really part of the same carbon cycle. Each can influence the others in myriad ways. Seen from this perspective, it's easy to understand how those with vested interests in maintaining the status quo have been able to confuse the public and some of our political leaders about the issue. After all, it's counterintuitive to think that global warming could cause more severe rainstorms, for example, or more snow in some areas, but it can. That there are so many different sources and sinks of carbon dioxide, the principal greenhouse gas, as well as other "climate forcing mechanisms" such as aerosols and other gases at play only adds to the confusion. Add to this a public that actually prefers to have a little warmer weather, and the appeal of the climate change skeptics becomes pretty clear. Why not wait just a little longer and see what happens?

As this series of essays shows, the scientific community has been saying for a very long time that the problem is clear and waiting will only make it worse. Yet those who want to maintain the status quo have thwarted them at every turn. Deny. Dismiss. Delay. It's the same general strategy used by the tobacco industry and others to obfuscate issues and keep the public confused. First, deny the problem exists. Fund plenty of paid experts to overwhelm the media and provide "balance." This can, and has, gone on for years. When the scales begin to tip and that tactic starts to fail, switch to dismiss. Accept that the problem exists, but insist either that nothing can be done about it or that changing our behaviors now will be more expensive than dealing with the unknown effects of the problem later on. Finally, when that tactic starts to fail, switch to delay. Accept that there is a problem that we need to deal with, but find ways to make it seem like we're doing something when actually we are not.

Accepting that our actions—the lives and activities of us puny humans—have fundamentally altered something as big as the atmosphere and the Earth itself requires a strikingly different way of thinking about the world and our place in it. Not that long ago our species consisted of little more than a bunch of rag-tag nomadic mammals, causing some localized but generally manageable havoc within Earth's systems. Our rapid population growth, combined with science and technology and a global economy, has forever changed all that. We are now a dominant force with the power to alter the biological and geological systems of our planet. For some, this might represent the apex of power for a unique and wondrous species. For others, it might represent the downfall of an arrogant creature, the only one known in the universe that seems intent on engineering its own demise.

One thing is absolutely clear: as the hand of god is more and more pulled by human strings, we have to accept responsibility for our actions because this particular god does indeed play dice with the universe—and right now it's on a losing streak.

.   .   .

PERCEIVED BENEFITS OF WARMING DON'T PAN OUT

Stuck out in the cold on freezing February mornings, those who live in northern climates can be forgiven for thinking "Bring on global warming!" But research tells us we should be careful what we wish for.

Wishful thinking about global warming isn't limited to cold northerners. In fact, some people have suggested that a warmer planet would be beneficial for humanity—by allowing farmers to plant crops in areas that are otherwise too cold, for example, and by increasing tree growth and creating more lumber. In theory, all that extra plant growth would also suck up carbon from the air, which could slow and eventually reverse global warming—handily solving the problem for us.

Sounds too good to be true, and it most likely is. However, it has been difficult to find out how plants will actually respond to prolonged increased temperatures. Experiments have been conducted on small plots using heat lamps, but these were very limited in scope, and they hardly mimic changes that would take place on much larger scales.

Scientists tell us we can expect more extreme weather events like heat waves and droughts as climate change progresses. Europe's summer of 2003 gave scientists the opportunity to examine what prolonged hot temperatures will mean for plant growth across a large area. That summer, temperatures soared across Europe, with averages exceeding the norm by six degrees Celsius. Rainfall also decreased by 50 percent compared to the average. It was a scorcher, and thousands of Parisians died from the heat.

Europe is fortunate to have an extensive network of scientific monitoring stations, giving scientists access to huge amounts of data. So the European Union commissioned scientists to mine those data to find out how the heat wave affected plant growth and carbon dioxide levels. The results were published in 2005 in *Nature*.

Researchers from seventeen countries examined crop-yield information and satellite data, along with carbon dioxide readings from fourteen forest sites and one grassland site. They found that Europe lost 30 percent of its plant life over the summer of 2003. This decrease in biomass (weight of living matter) combined with an increase in plant respiration (which releases carbon dioxide) means that, over the course of one summer, Europe's forests and fields released more carbon dioxide than all its plants had sucked up over the previous four years. Their findings do not include the release of carbon into the atmosphere from massive forest fires that also raged during that summer.

The report concludes, "In Europe, more frequent extreme drought events may counteract the effects of the anticipated mean warming and lengthening of the growing season and erode the health and productivity of ecosystems, reversing (carbon) sinks to sources and contributing to positive carbon-climate feedbacks." In other words, more droughts could actually speed up climate change and make the problem worse.

Of course, if temperature changes occur slowly enough, it's possible that forests and crops could acclimate and fare better. But those increased temperatures could also lead to other problems such as increased pests, diseases, and fires. The researchers say that more studies are needed to find out what to expect in the future.

We still have an opportunity to choose that future. By reducing the emissions that are causing the problem, we can slow climate change and reduce the threat. What's more, by becoming more efficient and less wasteful, the U.S. and Canada can become more economically competitive right now. Searching for ways to reduce pollution will also boost innovation and creativity, setting us up to be global leaders in the future.

Waiting longer to see what happens as our climate changes is not just wishful thinking—it's stupid. Climate change may be many things, but good for us is not likely to be one of them.

WARMING MAY CHANGE THE
NATURE OF THE FOOD WE EAT

Americans and Canadians are a well-fed bunch. They generally don't have to worry much about their food supply. For most North Americans, it's just a matter of heading to the nearest grocery store. But global warming and the need to move towards more sustainable ways of food production could gradually change what they eat and how they get it.

Most people have heard about the problems associated with global warming and what it will do to our climate. We are more likely to get longer periods of drought, for example, and heat waves could become more frequent or more intense. That could pose serious problems for North American farmers, especially on the prairies.

But if global warming also lengthens the growing season and warms more northern regions, it could have a beneficial impact on farming in colder climates—at least in some areas. Although more carbon dioxide (the main greenhouse gas) in the atmosphere from burning oil and gas is the primary culprit behind global warming, carbon dioxide itself can actually enhance plant growth. Commercial growers often exploit this enhanced growth by adding carbon dioxide to the air inside their greenhouses.

Because of these benefits some people believe that, although global warming will force changes to where and how North Americans farm, it might provide an overall net benefit to agriculture in Canada and parts of the U.S. The growing season is so short in cold climates that warmer temperatures and higher levels of carbon dioxide likely couldn't help but increase yields. Could this really be an upside to global warming?

Unfortunately, the issue is not that simple. Not surprisingly, nature is often far more complex than we first anticipate, and that's certainly the case with how plants respond to changes such as increased greenhouse gases in our atmosphere. For example, according to a 2007 article in *Nature,* very little is known about what other effects enhanced carbon dioxide levels will have on food. And some scientists are concerned that this knowledge gap isn't being addressed quickly enough.

As it turns out, higher carbon dioxide levels have other effects on plants, and not all of them are good. Many crops won't just grow faster in an enhanced carbon dioxide environment; they will grow differently. Generally, plants take up nitrogen from the soil in order to create proteins needed to help convert atmospheric carbon dioxide into sugars. But at higher carbon dioxide levels this job gets easier, so plants create less protein and take up less nitrogen from the soil.

But if plants don't create as much protein, then they could become less nutritious—for humans as well as everything else that eats them. This could have implications throughout the food chain, because many creatures depend entirely on plant-based proteins—including important livestock like cattle. Studies done on plants raised with higher levels of carbon dioxide confirm that they do indeed contain less protein, though scientists are not sure how serious the problem will be. Adding more nitrogen fertilizers to the soil could potentially make up some of the protein deficiency, but that poses other environmental problems, as nitrogen runoff from farms is already a major source of water pollution.

In higher-carbon-dioxide environments, the type of protein produced by plants also changes, which could alter the nature of some of our foods. Most types of bread, for example, depend on a specific kind of protein called gluten, which is key to making bread rise. Other foods could be affected too. Andreas Fangmeier, a German professor of plant ecology and ecotoxicology, once said that by 2050, carbon dioxide concentrations could make french fries poisonous, beer foamless, and wheat flour unbakeable. An exaggeration, most likely, but he raises an interesting point—one that we currently know very little about.

Global warming is a very serious problem. But it is also both fascinating and perplexing in its complexity. When everything is connected, you never know what one change in the natural world will mean to the entire system. We just have to remember that ultimately whatever changes we make will come back to us in the end. We had better choose carefully.

. . .

WARMER WORLD MORE ANNOYING, SCIENTISTS PREDICT

It's one of those science stories that at first appears rather irrelevant: a study out of the Woods Hole Marine Biological Laboratory found that poison ivy will become more common if our world continues to heat up from global warming. But the study actually gives us a hint about what a warmer world could really be like—and it won't be springtime in Paris.

Research has shown that plant growth tends to increase under higher carbon dioxide levels. So for six years, researchers at Woods Hole pumped extra carbon dioxide into three test areas of pine forest in North Carolina. By adding more of the most common greenhouse gas to the test forest, researchers hoped to simulate conditions we're expected to see in our atmosphere by the middle of this century.

Their findings, published in the prestigious *Proceedings of the National Academy of Sciences,* reported that when carbon dioxide levels for the experiment were raised by about 50 percent, poison ivy growth more than doubled—far exceeding average plant increases in the test areas. This is because vines respond especially well to increased carbon dioxide levels. Instead of storing the carbon in a woody stalk or trunk, as plants like trees do, vines simply grow more shoots and leaves, which provide them with more access to sunlight for photosynthesis—so they grow more quickly.

Interestingly, the poison ivy not only grew twice as fast, it became more poisonous. Researchers say it is unclear why this occurs, but it may have something to do with the way the ivy produces the most noxious form of urushiol, the vine's poison that irritates our skin. Another study, published in 2007 in the journal *Weed Science*, also found that enhanced carbon dioxide conditions can double the growth rate of poison ivy, as well as make the plant hardier. Other studies have found that global warming could increase common irritants as well—such as pollen levels in the air—making life more miserable for hay-fever sufferers.

One can imagine the humorous headlines resulting from such studies: "Life in future more irritating, scientists say," or, "Global warming makes world more annoying," but there is more to the story. For example, increased vine growth in forests also has a climate feedback effect. As vines grow faster, they can choke out woody plants such as trees, which store far more carbon in their trunks. So, rather than soaking carbon out of the atmosphere and storing it, these forests could soon start turning it over rapidly and make global warming worse.

In fact, our climate contains a number of such "feedback loops," in which a change brought on by warmer weather causes another change, which exacerbates the problem. As frozen tundra in northern Canada and Alaska melts, for example, it could cause the soil to release methane, which is an extremely powerful greenhouse gas. This in turn could again make global warming worse.

Also in the North, if global warming reduces snow and ice cover, it will expose the darker earth below. Without the bright white snow to reflect some of the sun's radiation back into space, the ground will absorb more light and heat, again potentially leading to increased temperatures.

Our climate is a complex system, intimately connected to the entire biosphere and all life within it. Small changes here or there can have repercussions down the line. Seemingly minor alterations can have cascading and unintended consequences. It would be foolish to assume our climate will change slowly in a simple, linear fashion.

We hear so much about the most dramatic problems global warming is expected to cause, such as rising sea levels, extinction of species, and more frequent or extreme weather events, that it's easy to dismiss less provocative studies. But they are all part of a story we need to understand if we are to prepare for a different world tomorrow and make the changes necessary to prevent some of those problems from occurring today.

EXTREME WEATHER EXTREMELY COSTLY

Global warming may have been the last thing on the minds of residents of Vancouver, British Columbia, as they dug out from a record snowfall and cold snap in November 2006. But it's another reminder of how much we all depend on the stability of our atmosphere.

While residents of other cities may scoff at Lotus Land's relatively minor misfortunes, Vancouver has certainly had its fair share of weather anomalies in the past few years. First, record rains churned up rivers and caused landslides in the city's watersheds, leading to turbidity problems in the drinking water supply and a boil-water advisory across the region. Then, just as the water began to clear, a record cold and snowfall paralyzed the city.

What has this got to do with global warming? Well, extreme weather events like these are exactly the kind of thing climatologists say will become more common as our climate heats up. How confusing is that? Global warming can cause heavy snowfalls. But it's true.

This ability to link global warming to so many weather-related phenomena has created a bit of a joke: blame everything on global warming. Stock market down? Global warming. Can't get a date? Global warming.

But underlying the joke is a serious fact. Our atmosphere is connected to everything—including us. By adding vast amounts of heat-trapping gases like carbon dioxide to the atmosphere (from our industries, cars, and power plants), we're trapping more heat near the surface of the Earth. More heat means more energy. Adding so much energy to our atmosphere creates the potential for more violent outbursts—like the weather Vancouver has been getting.

This is why it's so imperative and urgent for humanity to get this problem under control. It's not as though global warming is just a minor inconvenience. Left unchecked, it's set to become a major hindrance to economic growth and international development. Vancouver newspapers were full of stories during these extreme weather events about how much these "natural" disasters were going to cost the city's economy.

In developing countries, severe weather events are doing more than harming the economy—they're killing people. Of course, extreme weather has always killed people. But in an article from a December 2006 edition of *Science,* Indian researchers reported that extreme summer monsoon rains in India are becoming more common. In the summer of 2006, for example, more than one thousand people died during one torrential rainstorm around Mumbai.

For the *Science* study, researchers analyzed data from 1951 to 2000 recorded by more than 1,800 weather stations around central and eastern India. They found that while overall rainfall remained fairly consistent during the fifty-year period, the number of extreme rainfall events doubled. Researchers cannot conclusively say that human-induced global warming is the cause, but the study's findings are in line with what computer models predict will continue to happen unless we seriously curb greenhouse gas emissions.

This research helps shed light on why, when global warming models predict more rain in places like India, rainfall there doesn't seem to have increased overall. The answer is that although annual average rainfall hasn't necessarily increased, extreme rainfalls have. That's unfortunate because more steady rainfall could actually benefit India's agriculture. Extreme weather benefits no one, especially in a developing country like India that lacks the infrastructure to deal with it.

Keep that in mind for the U.S. and Canada. Many North Americans sure wouldn't mind more pleasant weather. But global warming won't benefit anyone if more extreme weather is the result. Just ask folks in Vancouver.

*   *   *

FOREST PROTECTION VITAL TO STEM GLOBAL WARMING
In 2007, when nations working under the Kyoto Protocol agreed that forest-rich tropical countries should receive carbon credits for protecting their forests, it opened up a whole new way to combat global warming.

Here's why: tropical countries have vast amounts of carbon stored in their forests—including the trees, the soil, and the peat. If that carbon ends up in the atmosphere as carbon dioxide, it will act as a heat-trapping blanket and greatly increase the growing burden of global warming—over and above problems caused by the burning of fossil fuels. It's already happening right now, with countries like Indonesia and Brazil leading the way in terms of emissions from deforestation.

When the Kyoto Protocol on global warming was drafted back in 1997 (yes, Kyoto is now more than ten years old, and we're still battling it out), forest protection, or "avoided deforestation," was specifically excluded as a measurable credit in reducing emissions. That's because forests can burn down or be otherwise compromised, so those emissions might go up into the atmosphere anyway.

However, evidence over the past decade has shown that forest destruction, in particular tropical deforestation, is a critical source of the heat-trapping gases that cause global warming. In fact, a paper published in 2007 in *Science* reported that up to 20 percent of human-produced greenhouse gas emissions throughout the 1990s came from logging in tropical forests. According to the paper, this trend is expected to continue unless we make a concerted effort to curb the destruction of tropical forests.

Questions over the permanency of tropical forests as carbon sinks are certainly legitimate. But answers to some of those questions have been found in the past ten years. For example, an early study found that business-as-usual increases in carbon dioxide levels would raise temperatures in the Amazon to the point where the forests would die off, releasing much of the carbon dioxide they were supposed to be storing. This study led many to believe that relying on tropical-forest conservation would be risky from a carbon-storage perspective.

But newer research has since found that the risks aren't as high as were originally feared. As of 2007, ten of the eleven studies done on the issue of tropical forests for the UN-sponsored Intergovernmental Panel on Climate Change (IPCC) concluded that tropical forests aren't so sensitive to temperature changes that they would start to die off and become sources of carbon in coming decades. This means that, although their ability to store carbon will gradually go down over time, we can expect tropical forests to continue to be carbon sinks right through this century.

Obviously, then, in addition to reducing greenhouse gas emissions from burning fossil fuels to combat global warming, we also need to avoid deforestation. According to the 2007 paper in *Science,* cutting tropical deforestation in half by the middle part of this century will reduce heat-trapping emissions by 12 percent of what will ultimately be necessary to keep our climate stable and avoid dangerous global warming.

Developing countries will be the hardest hit by global warming, as they do not have the infrastructure to deal with increasing extreme weather events, rising sea levels, and other effects of a changing climate. In many developing countries, deforestation is also the greatest source of greenhouse gas emissions. As a result, it's in the best interests of these nations to protect their forests. And it's in our best interests here in developed nations to help them, because ultimately, their fate is our own.

In 2008, the United Nations set up a framework that will allow developing countries to benefit from forest protection. The system, called Reducing Emissions from Deforestation and Degradation, or REDD, gives developing countries carbon credits for protecting their forests—credits that could one day be worth tens of billions of dollars. The program is in a trial phase and could become widespread by the next phase of the Kyoto Protocol, which begins in 2013. There's still much to be worked out over the next few years, but the program has tremendous potential to give developing nations a fighting chance in the ever-changing war to prevent dangerous climate change.

. . .

GREATEST CLIMATE CHALLENGE
MAY BE OVERCOMING IDEOLOGY

The other day, a friend asked if I ever got tired of writing about environmental issues. Well, to be frank, I'm getting pretty tired of writing about climate change. Not because it isn't interesting or relevant or important, but because I'm getting tired of having to defend the science against conspiracy theorists and ideologues.

Time and again it seems the public is burdened with yet another round of anti-global-warming conspiracy tales, punctuated with vague promises from political parties that add up to very little and even allow our greenhouse emissions to continue to rise.

I'm not the only one who's sick of this nonsense. A few years ago, Donald Kennedy, the former editor-in-chief of *Science,* the world's most widely distributed science journal, wrote in an editorial: "We're in the middle of a large uncontrolled experiment on the only planet we have." And he concluded with: "Our climate future is important and needs more attention than it's getting."

In 2004, to help gain that attention and dispel any lingering myths, the American Association for the Advancement of Science (AAAS) and the business-oriented Conference Board held a public forum on climate change. The two-day Washington, D.C., event included a panel of climate experts and Nobel laureates who discussed the current state of climate science and ways to help the public understand the urgency of the situation.

As expected, the panel members concluded that without a doubt, the world's climate is changing and that governments and consumers should take immediate steps to reduce the threat. The experts acknowledged that climate science isn't perfect and that questions remain, but they pointed out that current climate models are more likely to be too conservative rather than too generous with their predictions.

Harvard geochemistry professor Daniel Schrag told the panel: "We cannot wait for a catastrophe to appear before we act because by then it would be too late." And he pointed out: "This should not be a partisan issue." He's right, it shouldn't be. We're beyond that now. Instead, we should be discussing the most effective and innovative ways to meet Kyoto and become modern, efficient nations.

Even big business is recognizing the need to act. Four years ago, the chair of Shell Oil told BBC news, "No one can be comfortable at the prospect of continuing to pump out the amounts of carbon dioxide that we are at present." And *Toronto Star* business columnist David Crane noted that Lord John Browne, CEO of BP, one of the world's largest oil companies, wrote in the journal *Foreign Affairs* that climate change is a serious issue that must be addressed and that meeting the Kyoto Protocol will not be nearly as difficult as some industry groups claim.

If an oil baron like Lord Browne can acknowledge the importance of moving on this issue, and labor, medical, and religious groups all support the Kyoto Protocol, as do the majority of North Americans and the vast majority of scientists, why would anyone even entertain the notion of dropping out of the protocol? Countries would not only miss out on tremendous opportunities to become more efficient, modern nations, but they would look regressive and backwards on the world stage.

Yet that's exactly what they did. The United States backed out of the protocol first, followed by Canada a few years later. Canada's retreat from its promise was particularly alarming because, unlike the United States, the Canadian federal government had formally ratified the agreement, legally binding the country to meeting its greenhouse gas reduction targets. Canada's prime minister, Stephen Harper, has called the targets "onerous" and said that the country won't even try to meet them. Instead, as of 2008, Canada was still following a climate change plan that Al Gore called "a complete and total fraud." This does not bode well for a nation that is known around the world for its environmental ethic.

How many scientists must stand up and say, "Do something!" before our leaders take this issue seriously? The evidence of human-induced climate change is overwhelming. At this point, it's not only intellectually dishonest to claim we have no need to take action on climate change; it's morally reprehensible.

EXTREME WEATHER LIKELY TO INCREASE

After Hurricane Katrina plowed a swath of destruction through the southern U.S., most people were probably wondering how to help those in need or feeling saddened by the injuries and loss of life. Some people, however, were apparently more concerned with getting another message out: "It has nothing to do with global warming!"

Within days of the hurricane's strike, several Canadian newspapers published the same opinion article by the same author on how it has become "fashionable" for governments, environmental groups, and those in the media to blame extreme weather like Katrina on climate change. The author went to great pains to insist on no connection between the two.

Strange. I can't recall a single headline that read: "Hurricane Katrina hits U.S.—Global warming to blame" or remember a quote by an environmental group attributing the disaster to global warming. Fact is, newspaper editors didn't write those headlines, and scientists and environmental groups didn't say those quotes, because you can't attribute any individual weather event to climate change. It just doesn't work that way.

Certainly, some computer models suggest there will be an increase in hurricanes as a result of climate change in the future. And many computer models anticipate an increase in extreme weather in general this century, though not necessarily an increase in hurricanes. But the jury's out on whether such increases are already occurring.

Some studies, such as a paper published in a 2004 edition of *Geophysical Research Letters,* conclude that they are. It ends, "Thus our results suggest that predicted increases in Canadian forest fire occurrence due to anthropogenic climate change are already being observed." A paper in the *Journal of Climate* concludes: "In the midlatitudes, there is a widespread increase in the frequency of very heavy precipitation during the past fifty to one hundred years."

But other studies are inconclusive. Indeed, finding out if extreme weather events are actually increasing in either severity or in frequency around the world is difficult because of a lack of good-quality data from areas outside major population centers.

So why would the author send out this red herring addressing a non-issue? Because he may have other motives. He has been quoted in the press saying, "This (global warming) is the biggest scientific hoax being perpetrated on humanity. There is no global warming due to human anthropogenic activities."

Ah, so there you have it. Katrina, it seems, was just a convenient excuse to get the same tired "Global warming isn't happening, and if it is, it has nothing to do with anything people are doing" message out to the masses. The charitable among us might call that being opportunistic. The cynical would call it ambulance chasing.

In 2005, the world's most prestigious scientific bodies— the U.S. National Academy of Sciences, the Royal Society of the U.K., the Royal Society of Canada, and others—signed a declaration warning about the "clear and increasing" threat of climate change and urging our leaders to act. An analysis in *Science* of all 928 peer-reviewed climate studies published between 1993 and 2003 found that not a single one disagreed with the general scientific consensus on climate change. In 2007 and 2008, the national scientific academies representing the G8 plus five other nations followed up on their 2005 statement, again pleading with our leaders to take appropriate action and noting: "It is unequivocal that the climate is changing, and it is very likely that this is predominantly caused by the increasing human interference with the atmosphere. These changes will transform the environmental conditions on Earth unless countermeasures are taken." To ignore such evidence and insist on "proof" flies in the face of the way science actually works. Science does not progress in a direct, linear path. There are no straight lines from discovery to discovery to enlightenment. When I tell university students today about some of the ideas we had about genetics when I was their age, they burst out laughing. A 2005 analysis of scientific papers found that 50 percent of them are probably wrong. But that's not entirely unexpected. We learn from our failures as much as from our successes. That's the nature of the scientific process.

To demand absolute proof in science before acting on a threat is to ask the impossible. It's not just anti-scientific; it's anti-science.

.   .   .

CLIMATE CHANGE MYTHS DEBUNKED

Despite explosive news coverage about global warming over the past two years, most people still have only a very rudimentary knowledge of this complex issue. Unfortunately, this lack of knowledge has led to persistent myths, which are slowing down real action that could prevent the worst damage from occurring to our economy and to our environment.

Most of us are just too busy to get to the bottom of climate science. It's undeniably complicated, and it's more than most people want to deal with in their daily lives. We all have to worry about our jobs, our families, and just getting through hectic days. Global warming is scary, and we hope someone does something about it or tells us what to do.

For some, however, doubting the science of global warming has taken on an almost religious zeal. Those blessed with "knowledge" shake their heads sadly at people who are concerned about a warming planet and are trying to do something about it. They pontificate about how the public has been misled by a few (usually European) academics who rely on "faulty" computer models, socialist biases, or both.

Talking to these people is hard because they come armed with obscure-sounding references about things like the "medieval warm period," "solar flares," and "hockey-stick graphs." They seem so sure of themselves that the media still routinely feature these so-called global warming skeptics in opinion articles, television interviews, and especially on talk radio.

Media outlets love these guys (yes, they are mostly men, and they tend to be the same, often paid, "experts" over and over again), because they stir things up. These pundits specialize in arguing and confusing people, the same way tobacco industry lobbyists did and still do. Having people argue on talk radio is that medium's bread and butter. And what better way to get people riled up than to have a self-proclaimed "expert" tell everyone that global warming is a myth?

The problem is that some people believe it. Or, more often, it creates just enough doubt for people—including politicians—to ignore the issue. And that's dangerous.

Many environmental organizations' websites correct some of the most common myths perpetuated by climate change skeptics, but the best resource I've seen yet comes from a magazine. *New Scientist,* the world's largest general interest science publication, has a feature called "Climate change: A guide for the perplexed," and it debunks twenty-six of the most common myths about global warming. Available free online at www.newscientist.com, the guide is an invaluable resource for separating fact from fiction.

New Scientist journalists did an impressive job of sifting through the most common misconceptions about global warming, exploring everything from computer models and hockey-stick graphs to ice core samples and various temperature readings. They looked at what the best evidence indicates as well at what areas need further research. It's a fascinating piece of work.

And it's badly needed too. Many governments are still stalling on taking substantial steps to reduce the heat-trapping emissions that are causing global warming. As the scientific academies representing thirteen nations wrote in their 2005 joint statement on climate protection: "The problem is not yet insoluble, but becomes more difficult with each passing day."

That's why it's so important to debunk these myths and move on. They're slowing us down at a time when delay makes the problem more and more costly and more and more difficult to fix. If you want to help, read *New Scientist* and arm yourself with knowledge, then tell a friend or, even better, an elected leader, and take down these myths once and for all.

*   *   *

PUBLIC DOESN'T UNDERSTAND GLOBAL WARMING

Have you ever participated in a focus group? They're very odd. Often used in marketing research, these small selections of randomly chosen people are brought together as a sampling of public opinion to gauge how folks feel about a particular product or issue.

A couple of years ago, my foundation conducted a focus group about global warming to see where people were at in their understanding of this complex and challenging problem. The results? Let's just say they were disconcerting, to say the least.

Simply put, most people didn't have a clue. The majority felt that global warming was a very important problem and they were quite concerned about it. But when pressed as to why it was a problem or what caused the problem, all heck broke loose.

Apparently, according to the average Joe, global warming is happening because we've created a hole in the ozone layer, allowing the sun's rays to enter the atmosphere and heat up the Earth—or something like that. The cause of the problem is cars, or airplanes, or aerosol cans. No one really knows for sure.

This is really quite remarkable. I would have thought that such confused understandings of the issue would have been commonplace several years ago, but with global warming being in newspapers on practically a daily basis, on the front covers of magazines, and in theatres (*An Inconvenient Truth*)—and a hot political issue as well—surely people would get it by now.

Apparently I was wrong. People don't get it. This is a big problem, because if people don't get it, then they don't really care, so politicians and CEOs don't really care and status quo rules the day. And blindly we march into the sunset.

But while science magazines are all talking about carbon sequestration and climate-forcing mechanisms, the average person is still trying to decipher the nature of the problem itself. True, few citizens need to understand the complicated nuances of atmospheric science or the various mechanisms of the Kyoto Protocol, but people cannot care about things they do not understand. If our leaders are to take the issue seriously, the public must have at least a basic understanding of it.

So, to clarify—the ozone layer is a part of the atmosphere way up high that helps shield the Earth from the sun's most harmful rays. A couple of decades ago, scientists realized that some of the chemicals we were using in our industries and homes were finding their way into the upper atmosphere, reacting with the ozone and destroying it. Scientists were concerned that if this continued, it would thin the vital protective layer, leading to increased skin cancers and crop damage. They sounded the alarm bell, the international community responded with the Montreal Protocol to phase out ozone-depleting substances, and today the ozone layer is gradually healing itself.

Global warming is a quite different phenomenon. Again, it's a human-made problem, but this time it's a result of the heat-trapping gases we are putting into the atmosphere from our industries, cars, and homes. These gases trap more heat near the Earth's surface. More heat also means more energy in the atmosphere, which means more frequent or severe extreme weather events like droughts, storms, and floods.

With each new piece of research, the expected effects of global warming become clearer, more urgent, and more disturbing. Scientists say this will be one of the biggest challenges humanity will face this century. Right now we are not tackling the issue fast enough or directly enough to escape the most severe consequences.

So if you understand what global warming is and what it isn't, please tell your friends. Please speak up and help ensure that we don't continue to grope blindly into the future, searching in the darkness for a light switch. Because at this rate, by the time we finally reach it, it may no longer work.

*   *   *

PUBLIC INTEREST IN GLOBAL WARMING STILL HIGH

In early 2007, a friend told me that public opinion and media fascination with global warming would be over in six months at most because the public is fickle and the media are obsessed with the latest trends. My friend clearly forgot to inform the public and the media.

Over a year later, a quick scan of the "latest news" page posted on *New Scientist* online found no fewer than six stories about global warming: "Europe's recent heat waves aren't a mirage," "New flood warning to save rural Bangladeshis," "Revealed: America's most polluting power plants," "Sunshade for global warming could cause drought," "Asia's brown clouds heat the Himalayas," and "Early springs show Siberia is warming fast." A Google search for news stories about global warming found more than twenty-six thousand in October 2008 alone.

Many of these stories weren't just in the science news, either. Several, including the more obscure articles, made it into the mainstream press. I think that rather than getting bored with global warming, reporters and readers are surprised by how complex and interesting these issues really are.

Some of the stories are quite simple—like the one on heat waves. According to the latest research, Europe's recent heat waves are part of a trend that shows increasing numbers of very hot days on that continent. In fact, today there are three times as many very hot days in Europe every year as there were in 1880. Interesting. And pretty simple, really.

But other stories are decidedly more complex, making them harder to understand but much more fascinating in that they help explain how our planet works. Our atmosphere, for example, is complex and connected to everything else in the biosphere (that thin layer of our planet in which life exists—including the air, soil, and water). Because everything is connected, small changes in one area cause large, unexpected changes in another. And global trends and regional realities can actually be quite different.

Take the story on Asia's brown clouds. For years, brown clouds of pollution have wafted over Asia—sometimes making their way all the way across the Pacific Ocean to North America. These sooty clouds come from burning wood, dung, charcoal, and fossil fuels in Asian countries, particularly China and India.

While most people tend to think of air pollution as just dirty air, it is actually a complex soup of particles and gases that all have different effects. Some of those gases or particles may hurt our lungs, for example, while others, like carbon dioxide, don't cause direct damage, but build up in the atmosphere and heat up the planet. Others can do both.

One of the least understood factors that make up air pollution is the effect of small particles, sometimes called aerosols, and how they relate to global warming. When sunlight hits aerosol particles in the atmosphere, the light scatters. Some of that light and heat is reflected back into space. This reflectivity is why aerosols have generally been thought to be cooling agents. In fact, many scientists say that all the aerosol pollution in our atmosphere may be masking as much as 50 percent of the impact of increasing greenhouse gases.

But a study published in a 2007 edition of *Nature* showed that though aerosols may have an overall cooling effect, locally they can do quite the opposite. The study, headed by Veerabhadran Ramanathan of the Scripps Institution of Oceanography, used unmanned aircraft to fly into Asia's brown cloud and take measurements. The team found that because aerosols can hold solar energy as well as reflect it, these brown clouds of pollution actually increased solar heating of the local lower atmosphere by 50 percent. These findings may help explain why the Himalayan glaciers, which are in the path of these brown clouds, appear to be shrinking at an alarming rate.

They also show us that we obviously need to consider the entire mix of what we put into the atmosphere and not just greenhouse gases. Global warming is a very serious problem and one that we are only beginning to understand. But interesting, relevant, and important scientific research is actually making it into the mainstream press. And in the long term, an educated public will be one of the most powerful tools in the fight against the problem.

6

YOU CAN'T GET
THERE FROM **HERE**

Car culture and global transportation

PEOPLE, BY AND LARGE, love technology. They love their
TVs, their cell phones, their computers, and all sorts of
other gadgets. But they don't have relationships with them.

Not so with cars.

People don't just love their cars as useful pieces of technology; they develop emotional attachments to them. They talk
to them. They bond with them. Some people would rather
divorce their spouses than give up their cars. It's a love affair
that has gripped the developed world for decades and is starting to take off in the developing world too—in places like
China and India.

All of this, of course, is a big problem.

Cars (and by cars I largely mean passenger vehicles, but
also road transportation in general) are one of the most inefficient ways to move people that you can imagine. Your average
passenger vehicle weighs more than 3,000 pounds. Your

average commuter weighs 180 pounds. So 95 percent of the
energy needed in this equation to get from point A to point B
is to move the car rather than the passenger. And since most
commuters are alone in their cars, it doesn't get any better. In
fact, it gets worse. While cars have improved in many ways in
the past twenty years, the average fuel efficiency of the 2008
model fleet is virtually identical to what it was in 1988—only
today there are far more cars on the road.

As we examine in this series of essays, cars create a huge
health and environmental burden—one that car companies
have tried to ignore for decades as they fought attempts to leg-
islate tougher fuel-efficiency standards. But the problem with
cars doesn't end with the vehicles themselves; it extends to
where and how we live to accommodate them. In fact, our cities have been designed specifically for cars, with low-density
suburbs that continue to sprawl farther and farther from city
centers, making them difficult and expensive to service with
public transit and virtually impossible to traverse by foot or
pedal power. On the outskirts of these suburbs are strip malls
with big-box retail outlets—all designed for cars. Naturally,
once these shopping destinations are established, more suburbs are built even farther out to service them.

The result is predictable: more cars, more traffic congestion,
longer commutes, more smog, and more of the greenhouse gas
emissions that cause global warming.

So how do we break free from this cycle? On one hand, it's
really quite silly. After all, cars are just hunks of metal and
plastic, so for goodness' sake let's get over this childish obses-
sion. Surely we're beyond mere peacocks, strutting around,
hoping in vain that others will equate the cars we drive with
sexual prowess, social importance, or intelligence. Plenty of
urbanites in places like Manhattan, London, and Vancouver
don't own cars because they are able to walk, cycle, or take

transit to everywhere they need to go. In cities like these a car is just an expensive inconvenience.

On the other hand, good luck. Try shopping without a vehicle for weekly groceries for a family of four in most North American cities. Try to afford a home in the central areas of Vancouver, San Francisco, or Seattle, where you could indeed live without a car. Try to get the kids to soccer and swimming and music lessons on public transit or on foot. Yes, some people can do it and that's fantastic, but for many people, living without that hunk of metal and plastic would be very difficult indeed. So, how do we reconcile this love affair with the harsh reality of a small planet with a limited biosphere and six and a half billion mammals all climbing over each other like crabs in a bucket?

Well, we aren't going to stop people from loving their cars, but we can get them to drive less with a series of incentives and disincentives. To begin, we need to have cleaner cars. Much cleaner. There's no excuse for having gas-guzzlers on the road when we have the technology to make them so much more efficient. That will require political will because voluntary promises aren't worth the paper they're printed on. But efficient cars are only part of the solution. People need to drive less, especially as single occupants. A carbon tax that makes gas more expensive while making public transit cheaper and more convenient is one way to get people out of their cars as daily individual commuters. Congestion tolls are working in many urban areas in Europe. And insisting on denser, pedestrian- and bicycle-friendly communities will make us less dependent on cars to get to schools and shopping.

Cars, or some version of them, are here to stay. We can't just wish them away and, given their growing sales in developing countries, the problem will get worse before it gets better. That's why it's so important to push for changes to make better cars and to make us less car-dependent now. The infrastructure we build today will still be with us for decades to come. We need to start building for the future now, because otherwise we may not like the one we end up with when we get there.

.   .   .

CONSUMERS CAN DRIVE ENVIRONMENTAL CHANGE

In 2000, I bought my first new car in more than three decades. A new car! "How could you?" you may ask. Don't cars represent all that's wrong with our relationship with the Earth?

In many ways they do. So I ride my bike and walk as much as possible, but like many North Americans, I find there are times when I can't avoid driving. A fossil-fuel-vehicle-free society remains my goal, but as much as I wish for it, getting there won't happen overnight. We have to start by taking practical steps like reducing our use of cars and making the ones we do have more efficient.

I made my decision to buy a new car after taking a trip with my daughter. She and a group of five other young women cycled across Canada to raise awareness about air pollution and global warming. For one stretch in Ontario I escorted them, providing support with a hybrid-powered pace car.

The hybrid car is a remarkable technological achievement. It has both an electric and a gasoline motor. At low speeds, the car uses only the electric motor powered by batteries. At higher speeds, the efficient gasoline engine kicks in. A computer decides which engine to use and it does so effortlessly, with virtually no noticeable transition between the two. This technology doubles fuel efficiency during normal city driving—greatly reducing greenhouse gas emissions and other pollutants. And you never have to plug it in.

I got excited about the car. I even agreed to be in a press conference as the first person in the country to buy one. Some folks criticized me for doing so, suggesting that it made me look like a pitchman for the auto industry.

I decided to take my lumps. I paid full price for my car, and I do not endorse any vehicle or any company. But we have to acknowledge positive steps when they happen. So when I see a technology that actually helps solve environmental problems, I take notice. And I take pride in knowing that I can be part of the solution. Green technology will come faster and cheaper if more of us demand it and buy it!

I believe the auto industry should take a greater responsibility for its role in damaging our environment and health, as Bill Ford of Ford Motor Co. did during his tenure as CEO— although his actions didn't necessarily measure up to his words. Despite all the advances made in pollution control technology, average new vehicle fuel consumption in North America actually rose by 13 percent between 1987 and 2004. The reason? Gas-guzzling sport utility vehicles, vans, and trucks, the industry's biggest sellers. Until a late 2007 revision to the U.S. Corporate Average Fuel Economy regulations, these vehicles were exempt from stricter passenger-vehicle fuel-efficiency requirements.

All major automobile makers played a part in glamorizing SUVs. As a result, the vehicles became a status symbol, a seemingly natural progression in the evolution of the working person's career. For more than fifteen years it was accepted that, of course, the first thing you did when you earned more money was buy an expensive SUV. It was just what you did. Those who drove smaller vehicles that cost less than what they could afford were deemed eccentric.

Automakers liked SUVs because they were cheap to produce and sold for a premium. But by 2006, consumers were fed up with high gas prices and began to look for other options—options that included hybrids. I bought my hybrid early on, partly out of a desire to reduce my own carbon footprint, but also as a political statement. And I don't think I was the only one. I wanted to help take away the industry's excuse that in making huge SUVs, they were merely fulfilling market demand. By switching to more fuel-efficient vehicles, I thought we could send them a message that we don't want huge, heavy, gas-guzzling vehicles on our roads. We could show government that we wanted change, that fuel-efficiency requirements were too low. We could reduce the demand for oil, thus encouraging the oil industry to pursue other forms of energy. We could reduce air pollution and global warming. And we could save money! Potentially lots of it, given the high cost of fuel.

That was nearly a decade ago, and things have indeed changed. Almost every major automaker now makes some version of a hybrid or uses other technologies to achieve better fuel efficiency. And there are even better options waiting in the wings. Call me eccentric, but I still have my little hybrid, and I plan on keeping it until an even cleaner technology comes along.

.   .   .

GIVE YOUR HEAD A SHAKE
What do you do when seven hundred of the world's leading climate scientists representing more than one hundred countries release a disturbing new report on the impacts of global warming?

Well, if you're vehicle manufacturing giant Daimler-Chrysler, you launch a campaign to mass-market the world's largest SUV in North America. More cup holders for everyone!

That's what DaimlerChrysler actually did in 2001 when it announced that it would sell a giant vehicle, called the Unimog, through the corporation's heavy truck subsidiary, Freightliner. The Unimog was based on a German military transport, and it weighed in at more than eleven thousand pounds—that's twice as heavy as your average full-sized SUV. Naturally, it used more fuel, too, getting less than ten miles per gallon.

As I noted in a column following news reports of the giant vehicle's conception, the timing was so sad it was almost comical. Weeks before news of the Unimog broke, after reviewing studies from all over the world, scientists with the UN's Intergovernmental Panel on Climate Change (IPCC) told us that the impacts from global warming this century were expected to be extensive, and in some cases severe. Some countries could expect a reduction in crop yields and an increase in drought while others might experience an increase in flooding. Even if sea levels did not rise, the report warned, coastal flooding from storms could affect up to two hundred million people in the next eighty years. At least the Unimog would come in handy during a flood. It stood close to ten feet tall, and you had to climb a ladder to reach the cab.

The report also noted that natural systems like coral reefs and other sensitive areas "may undergo significant and irreversible damage" because of global warming. That warming, according to the conclusions of another international panel of scientists earlier in 2001, is mostly attributed to human activities, especially the burning of vast quantities of fossil fuels like diesel and gasoline.

Meanwhile, at an annual meeting of the American Association for the Advancement of Science (AAAS) that same year, delegates learned that the rapid erosion of glaciers all over the world was likely a sign of an already changing climate. Mount Kilimanjaro, for example, the tallest mountain in Africa, had seen its ice cap shrink by 82 percent since 1912. At that rate, the ice would disappear entirely by 2020.

All of this made it a strange time indeed to launch the world's largest sport utility vehicle. The dubious honor of world's largest SUV had previously gone to the Ford Excursion. When it was released, Ford tried to assuage customers' fears of being confronted by these huge vehicles on the street, saying that they were largely for corporate use on rough terrain. But the majority of the Excursions sold until their demise in 2005 were luxury models going to wealthy families. Daimler-Chrysler apparently saw no need to downplay the purpose of their SUV. Unimog marketing manager Bruce Barnes told the *New York Times* that "Moms will want to take it to the grocery store. It's a head-turning vehicle."

I said at the time that we must hope for the sake of our climate and health that one day soon this type of monstrosity would be considered a "head-shaking vehicle" instead. Eight years later, it finally is.

DANCE, DAIMLERCHRYSLER, DANCE

Shortly after my article about DaimlerChrysler's plans to sell their massive Unimog commercial vehicle to North American consumers appeared in select newspapers, I received several letters from readers.

One letter, from Oklahoma, claimed that I was an "eco-nazi" who was "hell-bent on destroying the American way of life." The second letter was far more insulting.

It was from the vice president of public and government affairs for DaimlerChrysler Canada Inc. He wrote to say I was grossly mistaken about his company's desire to market the nearly ten-foot-tall, eleven-thousand-pound Freightliner

Unimog to the public as an SUV. He insisted that this was never the case, chided me for taking DaimlerChrysler to task about the issue, and concluded by asking me to set the matter straight.

The VP raised a key issue. Vehicle emissions soared because of the SUV/truck craze, but he assured me: "The Unimog *is not* [his emphasis], *and was never intended to be,* a sport/utility vehicle." Instead, he said, it was a commercial truck suitable for plowing snow, fighting fires, and working farms. He went on to write: "It is not intended to compete in the mass sport/utility vehicle market in Canada, the U.S., or anywhere else."

Perhaps he had not seen Freightliner's promotional brochure, the one with the front-page headline: "You don't need roads when you can make your own." The one that pictured the Unimog prowling city streets at night and proclaimed: "Tough, rugged, and eminently civilized." And: "Wanting to conquer the great outdoors is simply not a good reason to give up leather and air conditioning."

I'm sure that's the first option farmers ask for—leather.

Perhaps he was not aware of the speech Jim Hebe, then Freightliner's chief executive, presented at the Great American Trucking Show in Dallas. A press release on the company website quoted him as saying, "The Unimog adds a new dimension to both the North American 4 × 4 truck and SUV markets." That press release has since been removed because, a spokesperson told the *New York Times,* "people were misunderstanding."

People were probably also "misunderstanding" the Article of the Month in DaimlerChrysler's web-based magazine, *TransGlobal.* It stated: "Freightliner also hopes to market a SportChassis® Version of the Unimog as an attractive and alternative leisure vehicle. This would present a real challenge to the 4 × 4 monster pickups, known in the U.S. as SUVs

(Sports Utility Vehicles). These are extremely popular and represent a massive market."

Now, we could debate semantics here as to what actually constitutes an SUV and what actually constitutes mass marketing, but I think the story is clear. Granted, DaimlerChrysler planned on selling just a few hundred Unimogs every year to customers looking for a huge vehicle to drive around the city and turn heads. But those numbers add up. And if the Unimog had proven to be a marketing success like the Hummer had, sales could well have increased into the tens of thousands.

The funny thing is, if DaimlerChrysler had dropped its plans to market the Unimog as an alternative to a car because of environmental concerns, it would have looked like a hero. The press release could have read: "DaimlerChrysler puts clean air, global warming worries ahead of market share."

But they didn't do that. Instead, they denied ever having intended to market the Unimog as an SUV-type vehicle, even though all the evidence suggests otherwise. Perhaps DaimlerChrysler dropped these plans and desperately tried to sweep evidence of them under the rug. Or perhaps they still planned on quietly marketing the massive vehicle in some other format. In a follow-up essay I wrote after receiving the letter from DaimlerChrysler, I said: "Let's watch and see how they dance next."

As it turns out, the idea of the Unimog as an SUV appears to have danced into the sunset. It never did become a mass-market vehicle. DaimlerChrysler has since split up again into separate entities, and the backlash against giant SUVs—for environmental and fuel-economy reasons—has caused sales of such vehicles to plummet.

It's about time.

A GROWING BACKLASH AGAINST WASTE

In the early '90s, the decade was dubbed "The decade of the environment." It turned out to be quite the opposite, but the first decade of the new millennium still has promise.

Back in the early 1990s, environmental concerns topped the polls. Everyone, it seemed, was worried about the impact of human beings on the biosphere. It was as though the collective inertia of decades of unchecked industrial growth had suddenly caught up to us. When we stopped to take a breather, we realized the consequences of all our actions—pollution, global warming, habitat loss, and species extinction to name a few. We saw that our species had the power to alter the very systems that sustain life on Earth. And it scared us.

Unfortunately, it didn't penetrate our lifestyles. We did make some strides—we phased out ozone-depleting chemicals and started recycling and reducing some air pollutants. The problem was, once we made a few changes, we reverted to assuming everything was fine. Corporations and governments all developed bureaucracies focusing on environmental problems, and we thought they were taking care of things for us.

That made us all feel better, but it blinded us to other growing problems. Many of the well-meaning people who were putting out their blue boxes once a week then walked across their driveways to get into massive SUVs to drive to work—alone. Most didn't even recognize this as an environmental problem. They felt they needed a vehicle, and SUVs were big and looked safe. SUVs also looked like they could take people out to the wilderness in complete comfort. Who wouldn't want that?

Automotive manufacturers took advantage of this fantasy by using beautiful natural imagery in their commercials and advertisements. And boy were they effective. By the late '90s, the majority of cars hitting the roads of North America

weren't cars at all, but "light trucks"—a government classification term used to describe everything from pickups to SUVs
and minivans.

This classification provided a convenient loophole for vehicle manufacturers. Until 2007, light trucks were exempt from
stricter "passenger vehicle" fuel-efficiency regulations, so they
could burn more fuel and pollute more. Why? Because in the
late 1970s automobile manufacturers lobbied for the measure,
originally as protection against more fuel-efficient competitors
from Europe and Japan.

At the time it didn't seem to be a big deal. Back then pickups were largely for farms and industry, and no one had heard
of SUVs. That would change as the auto industry found it
could make enormous profits by slapping a passenger-vehicle
type of body on a truck frame powered by an old-technology
gas-guzzling engine. Toss in a dozen cup holders and leather
seats and dealers couldn't keep them in the showrooms.

So a loophole combined with effective advertising created a fad that led to more than a decade of massive growth
in greenhouse gas emissions in the transportation sector. Air
pollution and smog, which had been slowly shrinking in the
late '80s thanks to better fuel-efficiency regulations for cars,
once again began choking cities. North America went back to
where it started. Or worse.

But in the last two years, there've finally been some
changes. Led by California, these antiquated fuel-efficiency
regulations have received much-needed updates. And the
charge wasn't just championed by environmentalists. Average
people realized how wasteful large SUVs were. The "backlash
against SUVs" made the front pages of USA Today and the New
York Times. People finally saw them as big, dirty safety hazards
rather than the escapist fantasy machines they were portrayed
as in advertising.

But if tastes were changing, were improved regulations necessary? Couldn't consumers just decide with their wallets? Some would certainly say yes, but I hardly think it would be prudent to leave the fate of our health and well-being to automotive fads. When gasoline was cheap, fuel economy wasn't a huge priority for the public, so manufacturers simply ignored it. As a result, it wasn't until 2006 that the new vehicle fleet average caught up to where it was in 1987! Better regulations to reduce greenhouse gases and smog are critical steps to slow climate change and keep our cities healthy. We simply can't afford to leave the future to the whims of automotive fads and loopholes.

. . .

WE'VE HEARD THIS STORY BEFORE

One night while I was watching television, an ad came on for an auto company boasting about the effectiveness of air bags and how it was one of the leading companies concerned with safety. I sat up. Wait, didn't the auto industry lobby fight air bags with everything they had? And why did this sound so familiar?

Years ago, we did an episode on the CBC television program *The Nature of Things* about air bags. This was before they were actually in any cars, but they had undergone years of testing. Ralph Nader was riveting in his interview as he explained how air bags were a cost-effective way of saving lives and reducing injuries. The response of the automobile industry—all-out opposition—shocked me, even though its own data showed that thousands of lives would be saved. Today, those same companies have the gall to boast of their safety achievements.

I'm old enough now to remember similar problems decades back. I remember when people began to press for better air standards in Sudbury, Ontario, at a time when the city resembled a moonscape. Inco, the metal company responsible for killing all the vegetation, thundered that reducing emissions would lead to bankruptcy and they would have to shut the plant down. Years later, when Inco was forced to accept higher standards mandated by government, they found ways to not only reduce emissions but actually make money with the residues they captured and the new emissions technology they could sell. In addition, they garnered huge PR benefits by boasting of the greenery returned to Sudbury thanks to the company's "ecological concerns."

And who could forget the tobacco industry? We all know about that industry's lying, deliberate deception, PR spins, junk science, and cover-ups about the health effects of smoking. In 2000, the World Health Organization (WHO) released a report documenting in detail how the industry had been secretly working to discredit WHO's efforts to reduce smoking and to educate citizens in the developing world about smoking's effect on health.

We need to remember the historical response of industry lobby groups to the need for change as we witness the outrageous tantrums by the oil industry lobby when it comes to reducing the greenhouse gas emissions that cause global warming. Let's look at it this way: the National Resources Defense Council estimates that air pollution, much of it coming from burning fossil fuels, prematurely kills about sixty-four thousand Americans a year, while Health Canada says that up to sixteen thousand Canadians also die from dirty air. All kinds of technologies could reduce that pollution immediately at acceptable economic costs, which would mean thousands of lives could be saved annually. By bellyaching, dissembling, and delaying, the oil industry and their supporters are essentially telling us that we should allow further deaths because corporate profits are more important.

Time after time, we have been faced with problems in dire need of solutions—from the effects of cigarette smoking to deaths in automobile accidents and global warming. And time after time, industry lobby groups have said that economic catastrophe would result if solutions were adopted. Well, guess what? After new standards were enacted to protect the public, the sun still rose and the economy still chugged along just fine. Tobacco continues to flourish, and the automobile industry posted record profits after the U.S. Clean Air Act was implemented.

Maximizing profits at the expense of human and environmental health is not a god-given right. Government's role is supposed to be to set standards that benefit all, not just a few powerful industries. Yet big-business lobby groups like the oil industry still spend millions of dollars a year on advertising campaigns to convince North Americans that reducing greenhouse gas emissions could plunge us back into the Dark Ages. Think about that when you see those ads. Better yet, think about air bags and tobacco.

.   .   .

INDUSTRY NEEDS A PUSH TO BUILD BETTER CARS

Each fall—a crisp chill in the air, frost on the ground—automakers roll out their latest, largest creations.

For years, big was in. Big everything. And sex. Sex everything. Meanwhile, the streets got more and more clogged with vehicles and our air got smoggier. Polls showed that consumers wanted better, cleaner cars. We had the technology. If only the automakers made the effort.

In the United States, there are now more cars than drivers. Yet fall newspapers are still filled with automotive supplements, some almost as large as the newspapers themselves. They detail every type of new vehicle you can imagine, along with advertising for the same.

These days, about half of those vehicles are SUVs, pickups, and minivans. Until 2007, such vehicles were classified by the government as "light trucks," which means they weren't required to meet the same fuel-efficiency standards as cars. So they were allowed by law to burn more gas and pollute more.

That was one of the inequities former Member of Parliament Jim Fulton (and then David Suzuki Foundation executive director) wanted to address in a speech to automakers and journalists at a 2003 conference called "Cars and Culture," held in Toronto. The symposium was supposed to be an examination of the role of automobiles in society.

Mr. Fulton got up and talked about how he was a driver too, but how as a society we drive too much. There are too many big, heavy, polluting vehicles on our roads. Entire suburbs are designed around car culture. It's making us fat. It's responsible for increasing health care costs. It's making our kids sick.

In short, Mr. Fulton said, it's a huge problem that needs to be fixed. When we eat up valuable farmland to build isolated, public transit–challenged suburbs and force people to drive, we increase the smog in our air and add to our waistlines. When we allow massive SUVs powered by big, inefficient engines to roll off dealer lots, we get stuck with those obsolete vehicles on our roads for a decade or more. These decisions will affect our children for generations to come.

In Europe, where fuel-efficiency standards were much higher, Mr. Fulton said, the trend was not towards bigger, heavier vehicles, but towards smaller, more efficient ones. Two-seat "smart" cars were all the rage, combining fun styling with super fuel efficiency. They could park practically anywhere, and they polluted less and reduced traffic congestion. But you couldn't buy them in North America. They didn't arrive in the U.S. until 2008.

It isn't like there hasn't been the opportunity to change. In Mr. Fulton's speech, he talked about how, when he was a Member of Parliament way back in 1981, he proudly voted for legislation that would have greatly improved fuel efficiency for all vehicles—including SUVs. That legislation passed, but the auto lobby fought it like mad, so it was never proclaimed. As a result, new vehicles sold in 2004 in North America on average actually got worse gas mileage than they did back in 1980!

It's not like the technology eluded us, either. Hybrids (which have both an electric and a gasoline motor) burn far less fuel, but few models are available, and they tend to be prohibitively expensive. Yet in the U.S., the Union of Concerned Scientists took an SUV and made three thousand dollars' worth of simple modifications to it that increased gas mileage by 71 percent. In Europe, fuel-efficiency standards have been rising for many years, while North America's have remained virtually stagnant. As a result, Europe has developed far more options for consumers when it comes to fuel economy.

And North America still hasn't caught up. A 2007 study done by the U.S.-based Civil Society Institute found that there continues to be a growing "fuel-efficient car gap" between Europe and North America. According to this group, between 2005 and 2007 the number of vehicle models available in Europe that could achieve a combined forty miles per gallon increased from 86 to 113. During the same time period, the number of such vehicles available in the U.S. dropped to just two.

In his 2003 speech, Mr. Fulton ended by saying that industry needed to work with environmental organizations and smart-growth groups if we want North American cities to be healthy and livable for future generations. The audience, largely automotive executives and journalists, clapped. He sat down.

Then someone else took the stage and started talking about cars and sex. Executives sat up. Journalists started taking notes. General Motors unveiled a 1,000-horsepower Cadillac. Young women in tight shirts tossed keys to audience members and encouraged them to go for test drives in the latest, most powerful creations.

Sadly, things don't seem to have changed all that much since then. Sales of big vehicles have dropped due to the price of gas, but North Americans still don't have a lot of more efficient options. Consumers may be ready for better cars, but the auto industry still seems stuck in the past.

.   .   .

SUBURBAN SPRAWL BAD FOR PEOPLE AND THE PLANET

Each fall, millions of children across North America go back to school. But most don't walk or ride their bikes. No, most of those kids are driven. And it's making them fat.

In fact, this suburban, car-centric society is partly responsible for the near-epidemic levels of obesity for all age groups in North America, according to reports in the *American Journal of Public Health* and the *American Journal of Health Promotion*.

Researchers developed a "sprawl index" to measure patterns of development in communities across the United States. Then they compared the levels of suburban sprawl with the health of two hundred thousand people living in those communities based on responses from a national health survey.

Their results show a startling correlation between sprawl and obesity. In fact, people living in the least dense, most sprawling communities were likely to weigh six pounds more than those living in the most dense, compact communities. There was also a strong relationship to chronic disease—those living in sprawling communities were found to be more likely to suffer from high blood pressure.

By designing decentralized suburbs with little pedestrian or bicycle access to schools, offices, shopping, and recreation, we have essentially engineered physical activity out of our daily lives. For many people living in the suburbs, the car is by far the most convenient transportation choice to daily destinations. Many suburbs lack safe bicycle or pedestrian routes. Some don't even have sidewalks!

So while more than 70 percent of their parents walked or rode bicycles to school, just 18 percent of today's schoolchildren in the U.S. do the same. In fact, more that 90 percent of all urban trips in the U.S. are by car, while just 6 percent are by foot or bike. Canadian cities are, on the whole, more bicycle- and pedestrian-friendly. In 2005, 19 percent of Canadians' urban trips were made by foot or bike, but they still pale in comparison to the Dutch and the Germans, who walk or bike to some 40 percent of their destinations. These European countries also have lower incidences of obesity and heart disease.

The emphasis on the car not only is making North Americans fatter but is causing other health problems, like asthma. More cars on roads means more air pollution, which can trigger asthma attacks and other respiratory problems. It also means more global warming, which is likely to lead to greater health problems in the future, from increased air pollution to extreme weather events and increased exposure to new diseases.

Governments want to promote healthy lifestyles and have budgets for public awareness campaigns to encourage people to be more active. But it's absurd and counterproductive for people to have to get into their cars and drive to the gym! According to the U.S. Centers for Disease Control, simply burning an extra one hundred calories per day, achievable by twenty minutes of walking, is enough exercise to help people lose weight and curb the disturbing obesity trend.

To buck that trend, people in the U.S., Canada, and other car-centric countries have to reengineer physical activity back into their lifestyles. That means redesigning cities to be denser and more compact, with better bike paths and pedestrian-friendly walkways. It means changing development patterns so that "going out" does not always involve walking into the garage, climbing into an SUV, clicking the door opener, and driving three miles. It means fewer strip malls with vast parking lots that are accessible only by car. It means making stairways more accessible and more attractive so people use them. It means widening sidewalks and narrowing roads.

More important, it means cleaner air, better health, reduced global warming, and a better quality of life.

FOCUS ON HYDROGEN ECONOMY HAS THINGS BACKWARDS

You can't miss them. "Fuel-cell test vehicle" is usually written in big, bold letters next to the logos of a dozen big corporate sponsors. Fuel-cell manufacturers and automakers are eager to show off their fancy multi-million-dollar prototypes of hydrogen-fuel-cell vehicles. The cars are technological marvels. They're cool. They're futuristic. And they're virtually useless—for now.

As much as "gee whiz" stories abound about new hydrogen-fuel-cell technologies, one can't help but wonder if we aren't getting a tad ahead of ourselves. While it's great to have these vehicles to show off, it would be much better if we had a way to produce hydrogen in sufficient quantities cleanly. Or a way to store the stuff. Or an infrastructure to move it around. Or any number of a host of other major hurdles we need to jump before we are able to reach the vaunted goal of a "hydrogen economy."

It would indeed be an incredible achievement. The problem with today's "carbon" economy is that it depends almost entirely on fossil fuels for energy. These fuels, such as oil, coal, and natural gas, are non-renewable resources. There's a finite amount of them on the planet and the most easily reached oil and gas reserves have already been exploited. It's getting more difficult and more expensive to find remaining reserves. In addition, burning these fuels (as well as extracting and transporting them) releases greenhouse gases and other pollutants into the atmosphere. We all know the results—air pollution and climate change, to name just two.

Hydrogen, however, is the most plentiful element in the universe. It's capable of producing more energy per unit than any other fuel. And releasing that energy from hydrogen creates zero pollution. No nasty smog-forming chemicals. No heat-trapping gases. It's almost too good to be true.

Unfortunately, right now it is. The technical hurdles necessary to push fuel-cell vehicles into mass production are daunting. Even more important, if these challenges were met tomorrow, there is currently no way to produce massive quantities of hydrogen cleanly. Hydrogen has to be removed from water through electrolysis or from natural gas through reformation. Both methods currently require substantial amounts of fossil fuel energy, which releases pollution and causes climate change. In short, the benefits of a hydrogen economy will be lost if we have to use fossil fuels to produce the stuff.

A hydrogen economy will require a massive transformation of our current energy system to become more efficient and to focus on renewable sources like wind, solar, microhydro, geo-thermal, and tidal power. Only when we have large quantities of clean electricity available will it make sense to start produc-ing hydrogen for vehicles. This shift to renewable energy will take decades, giving researchers plenty of time to overcome hydrogen-fuel-cell hurdles. In the meantime, our air will get cleaner, and our climate will start to stabilize.

So why aren't we doing it? Unlike hydrogen fuel cells, the technology for this transformation exists today. It is not a technical problem but a political one. Two researchers from Princeton University point this out in a 2004 edition of *Science,* arguing, "Humanity can solve the carbon and climate problem in the first half of this century simply by scaling up what we already know how to do."

In other words, we can shift to a clean economy before we perfect hydrogen fuel cells. In fact, taking major steps to a clean economy now is a necessary condition to have a hydro-gen economy in the future. So let's just do it. Delaying action will only make the challenge more difficult. As a 2004 *Science* editorial points out, "Postponing action on emission reduction is like refusing to take medication for a developing infection. It guarantees that greater costs will have to be paid later."

A hydrogen economy may well be our future, but right now we need to focus on the present. The technology to start the shift to a clean economy exists right now. There's nothing futuristic about it.

FILL 'ER UP WITH SWITCHGRASS

Not long ago, the question at the pump was always, "regular or unleaded?" Today, leaded gasoline isn't even an option in most developed countries. And with the need to drastically reduce our consumption of fossil fuels, the question of the future just might be "switchgrass or algae?"

Of course, I'm being somewhat facetious. In their raw forms you couldn't run your car on either. But both organisms have the potential to be made into biofuels such as ethanol or biodiesel. And that, if done in a careful and sustainable way, could greatly reduce the greenhouse gas emissions that cause global warming.

However, despite some of the hoopla about biofuels, we still have many obstacles to overcome. Yes, you can already get ethanol mixed with your gasoline or biodiesel mixed with your regular diesel in many North American cities. In fact, in countries like Brazil, gasoline is always blended with at least 20 percent ethanol, and you can easily get 100 percent ethanol for your car. So far, so good. But these biofuels have problems too.

A widespread adoption of biofuels such as biodiesel and ethanol could cause serious damage to the environment and provide few benefits if the crop used to make the fuel isn't chosen carefully. Corn, for example, is the largest source of ethanol in the United States, but it is one of the poorest choices for fuel because if you do a life-cycle analysis (looking at all the energy needed to make the stuff), the energy obtained from corn-based ethanol is only marginally better, or worse, than the energy you put into it. Plus, corn is heavily reliant on fertilizers and pesticides. And using a food crop to fuel our insatiable appetite for oil to drive our cars doesn't seem very smart. It could even create food shortages in some areas if farmers can make more money selling their crops for fuel rather than food.

Corn is thus a very poor choice as a biofuel. But there are plenty of other options. Canola does better in a life-cycle analysis, for example, and sugarcane—which is where Brazil gets its ethanol from—better still. However, sugarcane requires a hot climate, and there are concerns that displacing Brazilian subsistence farmers to grow sugarcane will push them into slashing and burning the rainforest for cropland. So all biofuels still have an environmental, economic, or social cost. If these fuels are to be sustainable, such costs need to be minimized.

One promising biofuel that scores well in preliminary studies is cellulosic ethanol made from switchgrass. According to results of a 2007 study published in *Proceedings of the National Academy of Sciences*, switchgrass grown and managed for biofuel can produce 500 percent more renewable energy than the energy it needs to be grown and processed.

For the study, researchers conducted field trials (the first for switchgrass) over five years on ten farms in the midwestern United States. Looking at all the production and management information from each farm, the researchers were able to estimate greenhouse gas emissions and net energy inputs to outputs. After a life-cycle analysis, the results were very positive: greenhouse gas emissions from switchgrass-derived cellulosic ethanol on the farms were 94 percent lower than if the energy had come from gasoline.

Another benefit of switchgrass, and part of the reason for its success in the trials, is that it is a native prairie grass that grows on agriculturally marginal land. This means that fewer chemical inputs are required to maintain the crop, which makes it less likely that growing large crops of switchgrass would take away land that would otherwise be used for food production.

Biofuels have the potential to help reduce pollution and global warming emissions, as well as the regional conflicts caused by our dependence on fossil fuels. But choosing the right fuel crop for the right geographic area is critical, as is making sure that all social and environmental factors are considered. If we can overcome those hurdles, you can look for more biofuels made from waste wood, used vegetable oil, and yes, even algae at our pumps in the future.

CAR INDUSTRY HAS A HISTORY OF DECEIT

In the past, I've criticized the Canadian Liberal government for ignoring the best interests of its citizens and refusing to stand up to a corporate dinosaur. That dinosaur was the automotive industry, and the deal announced in 2005 to reduce emissions amounted to nothing more than another government subsidy to prop up an industry that feeds off the health and pocketbooks of Canadians.

The 2005 agreement, wrapped in obfuscating language about megatons of greenhouse gas emissions and navel-gazing preamble after preamble, didn't guarantee any improvements in gas mileage at all. In fact, it amounted to little more than a PR exercise for automakers and the feds.

On top of everything, it actually lied. Just look at the many preambles: "And whereas the Canadian Automotive Industry has shown good faith in meeting their commitments in other Memoranda of Understanding and are currently parties to numerous successful active agreements . . ."

A lie. Way back in 1982, the industry fought like mad to derail fuel-efficiency regulations that had already been passed by Parliament. Government caved in to the industry, which promised to meet the targets "voluntarily." On paper, they met those targets. But in reality, they avoided them by exploiting a loophole that allowed them to build and promote more and more gas-guzzling light trucks, which were exempt from the standard. The industry has also argued, threatened, and whined about the "impossibility" of everything from smog-reducing catalytic converters to safety innovations like seat belts and air bags.·

Here's another good one: "And whereas the Government of Canada acknowledges that the Canadian Automotive Industry has made significant progress in reducing greenhouse gas emissions and improving fuel efficiency since 1990 . . . "

Another lie. Fuel efficiency had decreased in the years leading up to the agreement because the industry was hell-bent on selling bigger, heavier, and more expensive vehicles like SUVs that cost less to make. But you don't have to take my word for it. Just look at a comprehensive analysis by economists Roger Bezdek and Robert Wendling published in *American Scientist*.

Bezdek and Wendling pointed out that average fuel economy for all new vehicles declined from 26.2 miles per gallon (8.9 liters per 100 kilometers) in 1987 to 24.7 miles per gallon (9.5 liters per 100 kilometers) in 2004. They then pointed to more than two dozen technologies identified by the U.S. National Research Council as technically feasible ways to make cars more fuel efficient. Then they examined the costs of using these technologies and what it would mean for the industry and the economy. Their input-output analysis concluded that making cars burn less gas would actually create more jobs, save money for consumers, and improve the economy—in addition to reducing smog and climate change.

Good for consumers, good for jobs, good for health—this win should have been relatively easy for the federal government. Polls showed that 90 percent of Canadians want more efficient cars and the vast majority don't believe the industry's lies that making them would be either too hard or too expensive. California already set the precedent in North America by enacting laws that require a reduction in greenhouse gas emissions from new cars, thus forcing them to improve gas mileage.

Instead, Canada got a voluntary agreement that guaranteed nothing and again left enough loopholes to drive a Hummer through. Canadians were being played for dupes by the industry, and the feds were letting them get away with it.

Unfortunately, a change in government hasn't helped all that much. Despite a huge rise in eco-consciousness between 2005 and 2008, Canada's Conservative federal government hasn't done much better. They crafted a new deal with automakers parroting the new U.S. national guidelines that were passed in late 2007. But these guidelines are again based on the lowest common denominator. California had already enacted tougher regulations, and these regulations have the support of many other American states, as well as the Canadian provinces of British Columbia and Quebec.

Once again, auto industry lobbying has won the day over the best interests of the people and the planet. And once again, federal governments haven't had the guts to stand up to them.

7

JELLYFISH—IT'S
WHAT'S FOR **DINNER**

Feeding the planet in the twenty-first century

TO FULLY APPRECIATE HOW screwed up our relationship with food really is, consider this: the children of North America today may be the first generation in two hundred years who, despite increasing medical advances, probably won't live as long as their parents. Why? They eat crap. And that's just one part of our increasingly dysfunctional food system.

Obesity, anorexia, malnutrition, smart food, fast food, microwave food, genetically modified food, cancer-fighting food—we are a society obsessed with how and what we eat. But for all that talk, most of us know very little about our food. In fact, our food has fundamentally changed in the past century. Humans have gone from being largely rural creatures—eating a limited diet and living close to the sources of our food—to an urban superspecies—having access to a wider diversity of foods but being isolated from the natural world and cut off from the origins of our own food supply.

Much of this transition was made possible by the green

revolution of the mid-twentieth century, which brought about a massive increase in food production. As a result, reduced food costs to the consumer enabled most people in the developed world to eat better than their grandparents could have imagined. At the same time, speedy and inexpensive global transportation gave us access to a wide variety of fruits and vegetables, even out of season. And new technologies enabled us to plumb the ocean depths in search of new sources of protein. Today, most of us have at least the potential to lead longer, healthier lives, in part because of this bountiful availability of food.

But these changes also have a dark side. In fact, our relationship with the food we eat—one of the most intimate relationships with other organisms that one can imagine—is now decidedly dysfunctional. In the media, the focus of this dysfunction is largely on consumption. That's not surprising since what we eat is so highly visible—both in our communities and in our bodies. This new relationship with food is evident all around us. Fast-food restaurants are as ubiquitous in our cities and towns as prepackaged, processed foods are in our grocery stores. And the amount of creativity that goes into inventing, creating, and marketing all these products is truly stunning—as has been the resulting effect on our expanding waistlines. Not to mention the frightening possibility that our children's lives are being shortened because of it.

But there's another side to our food that doesn't get nearly as much press even though it's actually more important—food production. Because most of us are now city dwellers, the origins of our food have become a bit of a mystery. Where does that shrink-wrapped chub of ground beef come from? What about that fish or that broccoli? And that frozen dinner or canned soup? They're often made from ingredients that come from all over the world, which are then taken to yet another location to be processed and then sent back out to stores, some of which are in the very communities the ingredients came from originally.

Most of us don't think about these things because we're completely removed from them. Food comes from the store, not from the farm. And even if it's been processed and reprocessed, it's still cheap. Can you recall a world without seventy-nine-cent hamburgers? If you're under the age of forty, probably not. The problem is, as with so many things, such prices don't reflect the true cost of these food products to the planet, to our society, and to our health.

As we discuss in this next series of essays, changing how we produce and consume food over the next few decades will be vital if we are to become a sustainable society. Although the international community has managed to control population growth somewhat, we are still expected to have close to nine billion people on the planet by 2050. That's a big family to feed, many members of which come from impoverished countries that currently lack the infrastructure and resources to sustainably manage their own food supplies. Yet in developed countries that do have such knowledge and capacity, agriculture and meat production have often become commercialized to the point that the bottom line is always the dollar and not the health of consumers or the planet.

This relationship isn't working. Food should be something to be enjoyed, shared, and cherished. It should be something we actively make a part of our lives, not sandwich in between work and what's on TV that night. Food production and food consumption may seem like two very different issues, but they are really two sides of the same coin. Developing a sustainable food system means rethinking both.

ORGANIC FARMING A GROWING FIELD

It wasn't long ago that organic farming was considered small potatoes—something practiced in rural areas by hippies and others looking to "get back to nature." Today, it is a large and growing industry, and you can find at least some organic food in virtually any supermarket. But is it best for our health and the environment?

In theory, organic methods of food production sound ideal. Strict rules prohibit the use of synthetic pesticides and herbicides, for example, which are also toxic to beneficial insects and plants and often toxic to humans. Organic farms also eschew the use of synthetic fertilizers because these nutrients can leach into rivers and lakes, boosting plant and algal growth, which can choke out other aquatic life.

The overall goal of organic farming is to use agricultural methods that have the smallest impact on the environment and offer the greatest benefits to people. That's vital, because feeding a growing human population will put increasing pressure on the natural world and society in the coming decades. In fact, a group of ecologists writing in *Science* argued that if we continue with current agricultural practices, the resulting damage "may rival climate change in environmental and social impacts."

But some critics say that organic farming actually increases pressure on the environment. In a 2001 opinion article in *Nature* titled "Urban myths of organic farming," a U.K. biologist wrote that organic agriculture is an ideology that results in "lower yields and inefficient use of land." The author further claimed that "although its supporters assert that organic agriculture is superior to other farming methods, the lack of scientific studies means that this claim cannot be substantiated."

That's a good point. Widespread use of organic agriculture in developed nations is a relatively new phenomenon, and until recently it did not receive much research funding—certainly not on the scale that conventional agriculture does. This lack of scientific evidence has created a bit of a backlash against organic agriculture. Several years ago an American television reporter even fabricated test results on organic produce and accosted organic producers at a trade show, insisting that their products were dangerous. He was later reprimanded by the network and he apologized, but these problems show the need for better scientific evidence.

A 2001 study published in *Nature* helped answer some of the critics' concerns. The six-year study, conducted at the University of Washington, found that Golden Delicious apples grown organically in an experimental plot ranked higher in terms of environmental sustainability, profitability, and energy efficiency than apples grown either conventionally or using a mixed method. In a taste test, untrained observers also rated the organic apples to be the sweetest of the bunch.

This study is especially important to developing sustainable agriculture because it considers many different variables, including energy use, impact on biodiversity, and soil quality, which all have an effect on natural resources. As the author of the study told *New Scientist* magazine, until this report, "there've been almost no studies looking at the overall sustainability of both [organic and conventional] methods." And the organic apple plots fared very well. For example, the soil held water better and resisted degradation, and the plots required less labor and less water per apple produced and provided similar yields.

Of course, one study does not tell the whole tale, but it was an important step towards developing a truly sustainable agricultural system. Since then, as we shall see, many other studies have supported its conclusions. Over time, we may find that no one system is ideally suited to all conditions. But the future of food production on an increasingly crowded and stressed planet must be based on sound science, not any particular ideology.

.   .   .

ORGANIC FARMING A REALISTIC ALTERNATIVE

Strange how a movement that began with the best of intentions has managed to generate so much animosity. I'm talking about organic farming. But while a few people seem convinced it's a scam, the research continues to suggest otherwise.

Organically grown food is certainly popular. People buy it for any number of reasons—they say it tastes better, they're concerned about the effects of pesticide residue on their families' health, and they believe it is less harmful to the environment. They're willing to pay a premium price for it too.

Because the modern organic movement is relatively new, there has not been a wealth of scientific data to confirm organic farmers' anecdotal observations that this method produces good yields while maintaining healthier soils and ecosystems. Such claims are too good to be true, according to some proponents of industrial agriculture. A few years ago, *The Nature of Things* did a program on organic farming. I thought it was a Mom-and-apple-pie-type show that everyone would love. To my amazement, we were inundated with letters of outrage from university agriculture facilities and chemical companies arguing that conventional monocultures with copious inputs of synthetic fertilizers, pesticides, and herbicides were the only way we could possibly feed our growing human population.

Today, some critics seem genuinely angry at the success of the organic movement. They've written books and published articles in journals saying that organic farmers are starry-eyed idealists who are trying to bring back nineteenth-century farming practices that will reduce yields by four times—and thus, if widely adopted, will lead to mass starvation.

But organic farming isn't about turning back the clock; it's about moving forward. It's about smart farming to help maintain healthy ecosystems. Conventional farming produces high yields, but there are also enormous costs—pollution of groundwater, rivers, lakes, and coastal areas and reduced soil productivity through nutrient leaching. The use of pesticides and herbicides also kills beneficial non-target species and poses a health risk to farm workers and potentially to consumers. None of these "external" costs is factored into the price of conventionally grown crops.

Organic farming seeks to reduce these external costs, and it seems to be working. According to a landmark twenty-one-year study published in *Science* in 2002, organic farming can produce good yields, save energy, maintain biodiversity, and keep soils healthy. The study took place on 1.5 hectares (3.7 acres) in Switzerland using four farming methods and several different crops. Crop yields, on average, were 20 percent lower using organic methods, but required 56 percent less energy per unit of yield. Organic plots also had 40 percent greater colonization by fungi that help plants absorb nutrients, three times as many earthworms, and twice as many pest-eating spiders.

Some crops fared better under organic systems than did others. Potatoes, for example, produced 38 percent lower yields, but winter wheat was just 10 percent lower. The researchers sum up, "We conclude that organically manured, legume-based crop rotations utilizing organic fertilizers from the farm itself are a realistic alternative to conventional farming systems."

Other studies have also yielded similar results. A paper published in the *Journal of Applied Ecology* found that using organic methods to grow tomatoes can promote biodiversity while maintaining productivity. A 2007 paper published in *Renewable Agriculture and Food Systems* compared thirty years' worth of crop harvests from both conventional and organic farms in the developing world and found them to have very similar harvests, but the organic farms had a much lower impact on ecosystems.

It is important to keep in mind that there is much that we don't know about agriculture and there is likely no ultimate answer to our food-production needs. To feed our growing population we have to be open to all ideas, new and old. And we mustn't let the entrenched interests of the commercial agriculture and biotechnology industries dictate the future of our food when less intensive and damaging alternatives are available.

*   *   *   *

INDUSTRIAL MEAT PRODUCTION
CAUSES MORE PROBLEMS THAN IT SOLVES

A massive scientific study that found high contamination levels in farmed salmon made headlines a few years ago, but the results shouldn't really have been surprising. Applying industrial production methods to raising food animals has caused problems at every turn.

Both beef production and salmon farming have been in the news in recent years—and not in a favorable light. According to a study published in *Science,* farmed salmon contains up to ten times as many contaminants such as PCBs and dioxins as wild salmon. Many of these contaminants are believed to be cancer-causing agents. In fact, contamination levels are high enough that the report authors recommend people eat no more than one serving of farmed salmon per month.

The salmon farming industry had criticized a previous study that found similar contamination levels because it used a small sample size, even though government agencies charged with testing for contaminants use similar sample sizes. This time, researchers tested some seven hundred fish—totalling more than two tons—from markets and wholesalers all over the world, including Los Angeles, New York, Vancouver, and Toronto. They consistently found high levels of contaminants in farmed fish.

The source of the contamination is likely the food fed to farmed salmon. Salmon chow is made from other, less profitable fish, harvested from all over the world. This fish is then ground up and made into fish meal. The problem is that chemical contaminants like PCBs "bioaccumulate" in the food chain through animal fat. This means that as one fish eats another, the contaminant concentrations get higher and higher. All wild fish, including salmon, suffer from this problem, but farmed fish fare the worst, probably because concentrating fish into meal accentuates the bioaccumulation process.

Proponents of fish farms will likely regard this as a minor setback for a growing industry. But salmon farming faces a host of other problems, from site pollution to sustainability issues. Right now, about six to nine pounds of wild fish are needed to be ground up into food to produce two pounds of farmed salmon. So we are depleting wild edible fish stocks to produce contaminated farmed fish.

Beef production has also been in and out of the news because of the discovery of BSE or "mad cow disease" in North America. While there have only been a few cases and the risk to humans of contracting the disease from eating beef is low, the incident again exposes the problems associated with mass-producing meat with a focus on profits rather than human or animal health and welfare. BSE would likely never have been a problem if factory farms did not try to decrease their food costs and increase the growth rate of cattle (and therefore profits) by feeding them meat products, including other cattle—essentially turning herbivores into carnivores.

Factory farming, whether it's for pork, beef, chicken, or salmon, treats animals like raw materials that are processed and turned into an end product—meat. Animals in these systems are literally treated like inert matter. Little thought is given to their welfare. They are crammed into small spaces, sometimes by the thousands, and fed antibiotics to increase growth rates and reduce infections. Salmon farms use pesticides to kill off parasites. Hog farms create so much waste that they pollute surrounding groundwater, rivers, and ocean shorelines.

By trying to force industrial factory–style processes that focus entirely on profit and efficiency onto the animals used in agriculture, we've created huge problems for ourselves. Is it really worth jeopardizing human health, polluting our water, and depleting our ocean fish stocks just to have ninety-nine-cent hamburgers, cheap pork rinds, and fresh salmon year-round? It's time to take a good hard look at our priorities and consider more than profits in the way we produce meat.

.   .   .

TOXINS IN FOOD SUPPLY SIGNAL NEED FOR CHANGE

"Higher levels of flame retardants found in farmed salmon," read the headlines in the summer of 2004. While the news raised questions about the safety of eating farmed fish, the bigger question might be—why the heck are flame retardants in salmon at all?

And farmed salmon aren't the only contaminated fish. According to figures released by the U.S. Environmental Protection Agency (EPA), one-third of all lakes in the U.S. and one-quarter of all rivers are contaminated with mercury or other pollutants to the point that people should not eat fish caught from them.

What's going on here? Well, it all goes back to a very simple but often forgotten point. In nature, everything is connected to everything else through air, water, food, and energy. So when we create chemical compounds that do not readily break down in nature, it should be no surprise that they start appearing in even the most far-flung corners of the Earth.

In the case of farmed salmon, the culprit chemical compound is polybrominated diphenyl ethers, or PBDEs, used widely as flame retardants in furniture and electronic goods. They are released into the environment when produced, but also through everyday wear and tear on the products that contain them.

The problem with PBDEs is that they are similar in structure to PCBs, which are linked to cancer and damage to the immune and reproductive systems. PBDEs are not as well studied, but they have also been linked to impaired learning and development. And while PCBs have been phased out and are slowly declining in the environment, PBDEs are increasing. In fact, levels of PBDE in human blood have doubled in less than a decade.

Persistent chemicals like PCBs and PBDEs disperse readily throughout the environment and tend to concentrate in the food chain through a process called bioaccumulation, so animals higher up the food chain tend to be more contaminated. That's the problem with salmon—especially farmed salmon.

Salmon are carnivores, and farmed salmon are fed concentrated oily pellets made from ground-up fish of other species. Wild salmon, meanwhile, tend to have a lower-fat diet of creatures lower on the food chain, so contaminants generally accumulate in them less readily.

Mercury is another toxin that can pose a significant health risk, especially to pregnant women and young children. The EPA survey that showed widespread mercury contamination in freshwater fish was nothing new. Surveys conducted since 1993 have consistently shown high levels of contamination in North America. Many commercial saltwater fish, such as swordfish and some tuna, can also contain dangerously high levels of mercury.

So with all these contamination concerns, should people just avoid eating fish? Not necessarily. Oily fish contain healthy omega-3 fatty acids. Instead of avoiding fish altogether, a better step would be to learn which fish are the least contaminated and which fish are grown or caught in a sustainable manner. Those species are a consumer's best bet.

But trying to avoid hazardous chemicals that are persistent and ubiquitous in the environment is not really an answer either. While contaminants like mercury and PBDE may concentrate in fish, they are actually found throughout the food chain. Eliminating them is the only real long-term answer.

So far, the European Union has banned two types of PBDEs, as have California, Washington, and Maine. In 2008 Canada finally banned these two types of PBDEs (which industry had largely phased out due to the bans in Europe and California) but failed to take action against a common, third type of the chemical. And although mercury emissions have fallen substantially in the last decade, this toxin, too, continues to persist in the environment. One of the key culprits is coal-fired power plants—which are also a major contributor to the heat-trapping gases that cause climate change.

Given the hazards of these chemicals and their persistence once released into the environment, governments should be seeking to eliminate them at the source, rather than putting the onus on citizens to seek out the least dangerous options in a contaminated world.

CURTAIL INDUSTRIAL FISHING
NOW BEFORE IT'S TOO LATE

One by one, population by population, we are pushing species to extinction. And if we don't take immediate action, the big ones could go the way of the dinosaurs, said the late Dr. Ransom Myers, who was a scientist with Dalhousie University in Halifax, Nova Scotia.

Dr. Myers was talking about fish—specifically the large predatory fish like tuna, marlin, swordfish, and cod that we prize for food. If we want to protect our ocean ecosystems, he said, we're going to have to catch a lot less of them.

A few years ago, Dr. Myers and Dr. Boris Worm completed a ten-year study of the world's fisheries, and the results were shocking. According to their report, which was a cover story in *Nature,* the amount of large predatory fish in the world's oceans has plummeted by 90 percent since 1950. In just fifty years, many populations have completely disappeared, while others are just barely hanging on.

Industrialized fishing is the culprit. Dr. Myers's report clearly shows that when large fishing boats arrive in an area, fish populations promptly collapse—usually by about 80 percent within ten to fifteen years. The rapidity of the collapse shows how brutally effective modern fishing techniques are and that decimated fish stocks do not necessarily rebound when fishing halts. Indeed, despite a decade and a half of closure, Canada's northern cod stocks off Newfoundland and Labrador—which have been reduced in size by 99 percent—have shown no signs of recovery since a moratorium on fishing them was imposed in 1992.

The report also indicates that many fisheries management strategies are fundamentally flawed. Most stock analyses were done years after the arrival of large fishing boats—which already may have decimated fish populations. So stock analyses are often based on populations that have already been reduced by 80 percent or more. This leads managers to set catch quotas as if depleted levels are normal, which means that stocks will never get a chance to recover. In fact, it will keep them at levels uncomfortably close to extinction.

Optimists might point out that at least with the top predators gone, prey species would flourish. They do—at first. According to the report, some flatfish and groundfish populations did initially increase when predators were removed, but those trends quickly reversed. Researchers surmise that prey species also began to decline, either because they were being incidentally caught along with the predator species, or because when predatory species started to disappear, fishermen began targeting prey species too. Fisheries scientist Dr. Daniel Pauly at the University of British Columbia calls this process "fishing down the food chain." At this rate, Dr. Pauly says, it won't be long before we're down to eating jellyfish for supper.

How bad things get in our oceans depends on how we react to emerging realities in the coming years. As one might expect, some representatives of the fishing industry have called the report "unnecessarily alarmist," as though a 90 percent reduction in fish shouldn't be considered alarming.

To their credit, both of Canada's national newspapers—despite being centered in Toronto, where fish are hardly a priority—ran relatively large stories about the report. In Seattle, where fish are obviously more important, the story was front-page banner headline news for the *Times*. Sadly, just up the coast in Vancouver, where I live, the story was pushed much farther back in one newspaper and didn't even appear in the other. Frankly, I think that's embarrassing.

Can we restore our once plentiful fisheries? Dr. Myers said it may be possible, but we have to act quickly. He recommended reducing fish mortality (through reduced catch quotas or reduced bycatch) by at least 50 percent. That will be a bitter pill to swallow for those who depend on fishing for a living. But if we don't do something soon, there may be no fish to be had.

.   .   .

AGRICULTURAL POLICY FOR THE BIRDS

How do you feed nine billion people? It's a daunting question, but one we really need to be asking ourselves if we hope to feed humanity without severely degrading the Earth's natural systems.

It may be hard to believe, but when I was born in 1936, there were just over two billion people in the world. In my lifetime, that number has tripled. Today, the United Nations' population estimates show that between now and 2050, another 2.9 billion souls will be added to our little planet. That's a lot of mouths to feed. But feeding them is just part of the challenge. Current intensive agricultural practices have a number of unwanted side effects—from pesticide use and fertilizer runoff, for example—that can harm wildlife, pollute water, and otherwise damage the natural systems that we too ultimately rely on for our health and well-being.

So the question really is, how do we sustainably feed nine billion people? A 2007 report published in *Science* provides us with some indication. As part of a comprehensive study, researchers with the University of Reading in the U.K. looked at bird population trends to develop a threat-based risk assessment model that will predict the impact of agricultural practices on biodiversity and ecosystem services.

Birds are especially relevant for such a study, because they can be very sensitive to agricultural practices. Populations of wild birds in the U.K. have plummeted by nearly half since 1970, and the government has committed to reversing the decline by 2020. Unfortunately, according to the assessment in *Science,* government policies designed to help the birds don't go far enough and, unless they are changed, bird populations will continue to decline.

Researchers developed their "crystal ball" assessment model by examining three basic needs for all birds: they all need a place to nest, they need a place to forage for food, and they need to be able to find food in their forage areas. The model also takes into account how vulnerable specific bird species are to changes in any of these areas. Some birds, for example, will nest only in a few specific types of bushes. If those bushes disappear, so do the birds.

To test their model, the researchers examined the major factors in which agriculture in the U.K. has changed and intensified over the past forty years. These include switching from spring to fall sowing, increasing chemical fertilizer use, the loss of wild natural areas, increasing drainage of the land, and increasing intensity of grassland management.

When they plugged these changes into their assessment model and looked at what it predicted would happen to fifty-seven bird species, the results strongly correlated with actual historical data. In both their modeled matrix and in reality, bird populations fell as these agricultural changes became increasingly more common across the country.

Next, researchers used their model to look at the future of sixty-two bird species. Much attention in the U.K. has been paid to the importance of conserving hedgerows to protect birds. Indeed, hedgerows are an important nesting habitat. However, when researchers used their model to predict the future of the bird species, the results were rather bleak. Most of them continued to decline, even if hedgerow conservation measures were successful. It turns out that the birds' future depends on what happens in the fields, not just the hedgerows—so changing farming practices will be essential to their survival.

Predicting the future of species using this kind of analysis is never going to be perfect, but it's an important and useful tool. As the researchers point out: "Unless the footprint of agriculture is carefully managed through sustainable development, both agricultural systems and remaining natural ecosystems will suffer further degradation, increasing the proportion of the world's species threatened with extinction and further limiting the ecosystem services they are capable of providing."

In other words, status quo isn't really a healthy option for humanity or the rest of nature—so we'd better use every tool at our disposal.

.   .   .

KEEPING AN EYE ON TRANSGENIC CROPS

Did you know that genetically modified or "transgenic" crops are now commonplace on North American farms? According to a survey in the United States, the majority of Americans have no idea just how pervasive this technology has become.

In fact, North Americans have been eating transgenic foods and using products made from their crops for over a decade. So, what kind of effect, for better or for worse, are these crops having on the environment?

One of the major concerns expressed by ecologists a decade ago was that transgenic organisms could inadvertently disrupt ecosystems by harming other organisms. Some transgenic crops, for example, have been engineered to resist certain types of herbicide. This allows farmers to liberally spray their fields with the herbicide, knowing it won't harm their target crop.

These concerns were apparently warranted, as farm-scale evaluations two years ago in the U.K. of some transgenic crops found that vigorous application of herbicides was also damaging to the diversity of other life forms around farms. That's because many of the weeds killed by the herbicides were important for butterflies and bees. Populations of these beneficial pollinators on the test farms fell, possibly having other, more wide-ranging implications up the food chain for birds and mammals.

Another common type of transgenic crop has an insecticide "built in." These crops have been genetically engineered to produce an insecticidal toxin that wards off pests. One of the most well-known has been engineered using a gene from a certain kind of bacterium called Bt. The advantage, in theory, is that crops carrying the Bt gene do not need to be sprayed with an insecticide to kill pests, and thus could be potentially cheaper and more environmentally friendly than their contemporary non-transgenic counterparts.

Concerns were raised, however, when lab tests showed that pollen from Bt crops was potentially harmful to non-target insects, making them grow more slowly or reproduce less often. However, a meta-analysis of the effects of Bt cotton and Bt maize on non-target insects in the field found that these types of crops appear, at least on the surface, to be less harmful to insects than farming methods that use insecticides.

This report, published in *Science,* looked at forty-two field experiments and found that fields of Bt cotton and maize contained more non-pest insects than did those that used insecticides to control pests. Of course, insecticide-free control fields still had the greatest number of insects overall. The authors point out that further studies to examine the impact on specific species of insects, rather than just all invertebrates, are essential to better understand the environmental impact of these crops.

Disturbingly, the researchers had to resort to obtaining much of their information on Bt crops through the U.S. Freedom of Information Act, because the companies that produced them did not publicly disclose it. The researchers also note that the debate around transgenic crops has been a heated and emotional one. "However, in the case of GM crops, scientific analyses have also been deficient. In particular, many experiments used to test the environmental safety of GM crops were poorly replicated, were of short duration, and/or assessed only a few of the possible response variables. Much could be learned and perhaps some debates settled if there were credible quantitative analyses of the numerous experiments that have contrasted the ecological impact of GM crops with those of control treatments involving non-GM varieties."

Transgenic crops are not simple products like widgets, iPods, or even automobiles. They are living organisms that can interact with other creatures in the environment in myriad ways. Nature is complicated. When you modify an organism at a genetic level, it shouldn't surprise anyone that the results are also complicated and often unexpected. Transgenic crops are, in many ways, radically new and should be subject to the greatest of scientific scrutiny, not suppressed by proprietary concerns.

. . .

ORGANIC FARMING STANDS THE TEST OF TIME

It might seem like a hip new trend, but organic agriculture has been the way we've farmed for nearly ten thousand years. Energy- and chemical-intensive, machine-driven industrial agriculture is a twentieth-century phenomenon. Although it practically disappeared in North America during the latter half of the twentieth century, organic farming has taken off again as both consumers and farmers have discovered the benefits of a more holistic approach to agriculture.

Organic farming is rooted in ancient knowledge painstakingly acquired and passed down through generations. Long before science could tell us why certain farming methods would produce greater crop yields, farmers were learning the hard way what worked and what didn't—and sharing their knowledge with others.

With the advent of industrial farming and the green revolution, this traditional farming was relegated to the status of "quaint" or "old-fashioned" in the industrialized world—something practiced by hippies on communes, certainly not by serious farmers. But while the green revolution initially produced higher crop yields, it also created new problems, from fertilizer and pesticide runoff to soil erosion and reduced soil fertility. Today, new studies are showing that organic agriculture can often match and sometimes exceed yields from conventional agriculture, conserving soil quality while eliminating the need for pesticides.

The Rodale Institute Farming Systems Trial is the longest-running comparison of organic and conventional farming in the United States. Since 1981, researchers have been planting crops at the Rodale Farm in Pennsylvania using conventional agriculture as well as two organic farming systems—one based on animal manure for fertilizer and the other based on nitrogen-fixing legumes.

In 2005 a review of the trial was published in *BioScience*. Researchers measured the economic feasibility of each farming system, along with its environmental impacts, energy consumption, and other indicators. They found that for some crops, like corn and soybeans, organic farming systems produced the same yields as conventional systems, but used 30 percent less energy, less water, and no pesticides.

In fact, during drought years, corn yields in the organic systems were 30 percent higher than those in the conventional system. Researchers say that the organic systems were able to perform better in drought conditions because their soils contained much larger amounts of carbon and organic matter. In the organic plots, this increased organic matter also led to an increase in the diversity of creatures, including twice the number of earthworms. In turn, increased diversity helped reduce damage from insect pests by introducing more natural predators.

One might expect the organic systems to have many beneficial environmental effects, but the researchers also found that the organic systems could be as profitable or more profitable than conventional systems. Although the organic systems required more labor (to remove weeds, for example, rather than spray them with an herbicide), consumers were willing to pay a premium for organics, so the profit margins were often better.

The researchers argued that organic technologies such as exploiting off-season crops, using more extended crop rotations, increasing the amount of organic matter in the soil, and improving natural biodiversity should be more widely adopted. They conclude: "Some or all of these technologies have the potential to increase the ecological, energetic, and economic sustainability of all agricultural cropping systems, not only organic systems."

In other words, many organic practices simply make sense, regardless of what overall agricultural system is used. Far from being a quaint throwback to an earlier time, organic agriculture is proving to be a serious contender in modern farming and a more environmentally sustainable system over the long term. With consumers expressing a preference for organics and farmers reaping the benefits, this is one trend that's likely to stay.

.   .   .

TRANSGENIC CANOLA NOT FRIENDLY TO BEES, BUTTERFLIES

While debate over the benefits and dangers of genetically modified crops has never achieved the level of intensity in North America as it has in Europe, studies are continuing on the effects these crops might have on the natural world. In 2005, the last of a series of British "farm-scale" evaluations was completed. And once again, genetically modified crops didn't come off very well.

The farm-scale evaluations conducted in England over the past several years are considered to be the world's largest ecological experiment on how new farming practices can affect nature. The first of these studies, published in 2003, looked at sugar beets, spring oilseed rape (canola), and maize (corn). It reported that, while transgenic corn fared better than its conventional counterpart in its effect on the environment, both the spring canola and the sugar beets fared worse.

For the 2005 study, researchers looked at conventional winter canola and compared it to its transgenic variety. In this case, the transgenic plants had been modified to resist a specific herbicide—allowing farmers to spray their fields liberally and kill weeds without harming their crop.

However, the herbicide is especially effective at killing broadleaf weeds, which are also preferred by bees and butterflies. Researchers found that as a result, butterfly populations dropped by up to two-thirds in the transgenic fields, and bee populations dropped by one-half. Some biologists are concerned about the long-term impact this could have on biodiversity in Britain and on creatures farther up the food chain, like birds, if the crop were grown on a large scale.

While these findings are certainly not a death knell for transgenic crops, they show that tinkering with the genes of an organism can have repercussions far beyond the "minor" modification intended by biotech scientists. That's because the scientists focus on controlled tests done in a lab or growth chamber. But those environments aren't like the real world where birds, insects, rain, and wind complicate things. The British study also shows that we must proceed cautiously with all transgenic crops. Every modification could have profound repercussions on the natural world and should be tested thoroughly, both in controlled situations and in field evaluations, before wide release.

Right now in North America, genetically modified crops are treated just like their conventional counterparts. They are considered "substantially equivalent" and thus not subject to any special regulations. In fact, the canola tested in the British studies has not been approved in that country but is already widely grown in Canada and the United States.

Companies that make transgenic plants insist that their crops are strictly monitored and have not caused any problems.

But Syngenta, one of the world's largest biotechnology companies, admitted in 2005 that it had for years accidentally been selling transgenic corn not approved for human consumption. More than 293 million pounds of the corn made it into the food chain in the U.S. That doesn't exactly inspire confidence.

Despite these sorts of problems, biotechnology is still considered the darling of modern agriculture. That's unfortunate, because we'd be better off looking at our entire food system rather than just a tiny part of it. In the farm-scale studies, for example, transgenic crops were tested against conventional counterparts using standard commercial agricultural practices. Yet we already know that these standard practices are causing problems as a result of soil erosion, nitrogen runoff, and pesticide use. It would have been even more interesting to compare the transgenic and conventional crops to an organic system, which has proven to produce similar yields without many of the problems associated with conventional agriculture.

Ultimately, what matters is that we develop a sustainable food system that provides us with healthy food and does not degrade the natural systems that sustain us and other life forms. Genetically modified crops may or may not eventually become part of that system, but right now they are being treated like our best bet. If our goal is to make money in the short term, then perhaps that's true. But if our goal is to create a safe, sustainable food system, the evidence just isn't there.

* * *

TRANSPORTING FOOD CAN COST THE EARTH

When it comes to food, buying local has been the mantra of environmental groups for years. After all, it's pretty easy to conclude that transporting fruits and vegetables from one side of the globe to the other isn't very good for the planet.

But a comprehensive analysis of the true costs of the way we produce, purchase, and consume food has found that while international transport of food does have an impact, when it comes to environmental damage, the big culprit is domestic transportation.

Researchers in the United Kingdom used data from previous studies to estimate the hidden costs of conventional agriculture in that country. These costs include things like government subsidies; exhaust pollution from transport trucks, railroads, and car travel; heat-trapping emissions that cause global warming; and infrastructure, such as roads.

Their results, published in the journal *Food Policy*, show that international ship and air travel currently contribute a relatively miniscule amount to the overall hidden costs of our food. By far, domestic transportation from the farm to the retailer and then from the shop to the consumer's home has the greatest impact—accounting for nearly half of the hidden costs.

Raw distance, it turns out, is not always the deciding factor in determining the adverse effects of transportation. Shipping by water, researchers note, has lower impact than shipping by road. Transport by air, however, has the greatest impact of all. Right now, hidden costs for the international transport of food are relatively low because much of this food is shipped by boat or in the cargo holds of passenger planes. If we start to ship food by air more often, these costs could increase dramatically.

But if domestic transportation costs in a country as small as the U.K. are high, then the hidden costs of food transportation in larger countries like Canada and the United States may be much higher. Consider a box of cereal, for example, which may start with wheat from the Midwest or the Canadian prairies that is transported to New York or Ontario for processing with other ingredients from all over the country, put into a box made in Maine or Quebec, and then transported to California or British Columbia for retail sales, where it will be picked up by a consumer driving an SUV.

Because of our reliance on fossil fuels for transportation needs, each of these stages has hidden costs. In fact, even buying local food but driving to the store to pick it up has increased hidden costs. So does this mean big-box chains that sell in huge quantities may unintentionally help the environment by reducing the number of trips taken to purchase groceries? According to the research, that doesn't appear to be the case. Consumers in the U.K. are actually making more grocery shopping trips and driving greater distances to make them than they were twenty years ago—before the rise of the megamart.

Another hidden cost of our food is taxpayer-funded government subsidies that prop up unsustainable agricultural practices. Switching to organic agriculture, the researchers conclude, would lead to big benefits in terms of overall costs to society. Of course, the benefits of organic agriculture in terms of environmental impact are greatly reduced if the food has to travel a great distance by road to reach the consumer.

So which food-shopping patterns will yield the most benefit to the environment and society? Looking at the data, walking, biking, or taking public transit to buy organic, locally grown (within fifteen miles) food would be the best choice. Grocery delivery services also help a great deal by reducing the overall number of vehicle trips. Even choosing a fuel-efficient vehicle and reducing the number of trips helps.

Unfortunately, suburban sprawl is rapidly eating up some of our planet's best farmland—which also happens to be located near urban centers. For our food to be sustainable, governments at all levels must work to curb sprawl and support local food systems.

RECONNECTING WITH FOOD IN THE SUMMER

Every summer, if I'm lucky, I get to spend some time with my family at our cabin on an island off Canada's west coast. It's a place we go to recharge our batteries and reconnect with each other and with the natural world.

Part of that reconnection is with food. Although many of us quickly scarf down whatever's convenient as we rush about our daily lives, eating food is one of the most intimate experiences we can have. The food we eat is broken down by our bodies at a molecular level and absorbed into our cells. It becomes part of us. We quite literally are what we eat.

That's why it disturbs me to see the kind of food many people consume on a daily basis. I admit, on occasion I'm guilty of less-healthy choices myself. I try to be vigilant about food, but I travel a fair bit, and it can be hard to find the time to slow down and eat right. People think that being on a TV show is glamorous, but after a long day of filming, my dinner might well consist of a veggie dog from the street vendor outside my hotel before turning in for the night.

When I get to the family cabin, food stops being a mere necessity to provide energy for another hour of shooting. It becomes something to celebrate. Summertime provides us with a bounty of fresh fruits and vegetables, and our oceans can still serve up a veritable feast of shellfish and other seafood. As the Coastal First Nations saying goes: "When the tide is out, the table is set."

Most of us in the Western world—in fact, the vast majority of us—now live in urban centers where we are often completely removed from the sources of our food. Much of what we buy is prepackaged, frozen, chopped-and-formed, or otherwise processed before we even pick it up from the nearest warehouse club store. So there's something truly special about digging up your own clams and mussels for dinner. Or catching a fish for breakfast. Or picking your own fruits and vegetables. Not only is the food fresh, the experience makes it taste better and feel more satisfying.

For the past twenty-seven years, part of my family's start-of-summer ritual has been to go cherry picking, because I wanted my children to celebrate food's seasonality. We stuff ourselves silly with the juicy red fruit and bring back pallets of cherries to share with friends. It's actually pretty hard work. But that's part of the fun and the satisfaction. You can't buy that experience from a big-box store.

In fact, it drives me nuts to go into a supermarket in the summer and see it loaded with imported fruits and vegetables when local gardens and farms are overflowing with food. Farmer's markets are where I prefer to get my produce in the summer, when local farmers and some industrious city gardeners make their harvests directly available to the rest of us.

There are plenty of reasons to support farmer's markets and local food besides the experience. Eating locally grown food helps reduce the pollution caused by transportation; apples from New Zealand, for example, are a pet peeve of mine. Many local farms often also have organic certification, which means they are less intensive and more sustainable in the long term—and organic produce is grown without using chemical pesticides.

Some proponents of organic food also say that it's better for you, although the research is inconclusive. A ten-year study published in the *Journal of Agricultural and Food Chemistry* found that levels of certain cancer-fighting antioxidant chemicals were almost twice as high in organic tomatoes as they were in conventionally grown tomatoes. Researchers surmise that the availability of nitrogen in the soil is the reason for the difference. But other studies on wheat and carrots have found little nutritional differences between conventional and organic crops.

Regardless of your reasons for eating locally, summer is a great time to slow down and reconnect with food. Few things are as fundamental to our personal health and well-being. And few things we do have a bigger impact on the health of the planet, either.

. . .

HOW MUCH IS A TOMATO WORTH?

In fall 2007 a column appeared in the *Vancouver Sun,* my local newspaper, about the trend of eating locally grown food. The author began by describing some municipal initiatives to encourage growing local food and then arrived at the thesis of his article: "The eat-locally, grow-your-own phenomenon isn't about access to affordable food, it's about smashing the capitalist system."

At first I thought it was some kind of joke. But the author went on to describe basic theories from Economics 101 like "comparative advantage" to show how nations that specialize in what they make most efficiently and then trade with other nations that also specialize in what they make most efficiently end up with more stuff than if they each made those same things on their own.

His point relating to local food was that most of us don't grow our own food because it's cheaper (or maybe he means easier, since theoretically you could grow food for close to free) to buy it from someone who can do it more efficiently than you. Thus, he concludes, "Buy local campaigns are attempts to disrupt international trade."

If this sounds nuts, that's because it is. Is the nice elderly lady down the street really thinking, "Screw the Chinese!" as she harvests fresh, tasty snap peas from her community garden? I hardly think so. More likely, she's getting healthy food and enjoying herself while growing it. And that's really the issue: our current economic system by and large completely ignores important facets of life that are worth a great deal but have never been assigned a monetary value.

Consider this sentence from the column: "The tomato you grow yourself may seem to taste better than store-bought, but it won't be cheaper." Note the word "seem," as though the tomato doesn't *actually* taste better, it only *seems* to—presumably because of the satisfaction you received from growing it. But even if that is the case, then you still enjoyed growing the tomato in the first place—and isn't that worth something? Why is it okay to put a dollar value on our labor, but not our pleasure?

And this is the problem. Only things that you can actually buy have a monetary value. So the value of a tomato is only what someone will pay for it. Not in the satisfaction of watching it grow, or the feel of the earth between your fingers when you plant it, or the warmth of the juice from the summer-ripened fruit when you bite down on it. None of these things has value because you can't buy or sell them.

Another thing that isn't valued in our economic system is nature—more specifically, natural services like cleaning our air and water and providing a stable climate. Things grown halfway around the world and flown to our doorsteps get a lot more expensive if you actually include the cost of the damage this does to our atmosphere. So we cannot know the real price of our food unless we do full-cost accounting, which considers all of these factors that traditional economics considers "externalities." Even then, we still haven't factored in the value of community, of spending time outdoors with friends and family, and so on, that you might get while growing your own food. What are these things worth?

Needless to say, the article had me pretty depressed. Was this really how people thought? But then an uplifting thing happened. I picked up the newspaper a couple of days later and there they were—letters. A whole page of them, in fact, from people who thought the original column was nuts too. Each of them pointed out various flaws, but all got at the same thing: our economy is a social construct that depends on the environment and our values, not the other way around.

Reading those letters gave me hope. People get it. And more and more of them are getting it every day. Obviously, we still have a long way to go as a society, but simplistic economics that devalue some of the most important things in life are finally going the way of the dinosaur. And that's as it should be, because human life does not begin and end with a dollar sign.

8

THE TRUE COST
OF **GADGETS**

Technology and the culture of consumerism

IMAGINE IF YOU DECIDED to throw away your cell phone, close down your Facebook account, disconnect your high-speed internet modem, unplug your satellite television receiver, put away your BlackBerry, shut down your iPod, turn off your DVD player, and abandon your HDTV. Friends might think you've lost it. Family members might suggest counseling. *"What's wrong?"* they would want to know.

And you could tell them you're leading a completely modern life, circa 1995.

That's right: all of these things that are so ubiquitous today either had yet to be invented or were relatively rare just over a dozen years ago. Today, they're must-haves. And while it's true that you could play this same thought game for virtually any time period in the past few hundred years, never have there been so many high-tech products available on the market, never have they been so ingrained in our lives, and never have they changed so quickly. Indeed, at this rate, by 2020 my dozen-year thought exercise could shrink to just five. After all, that DVD player I mentioned—a product that didn't even hit the market until 1997—is already obsolete and is being replaced by a new high-definition model.

Of course, the problem here is twofold. First, while some of these new technologies can reduce our environmental footprint by consuming less energy, for example, or by using less harmful chemicals in their manufacture, the sheer volume of their production soon overwhelms any environmental benefit. From the constant need to upgrade to the latest model to the marketing collateral and the outrageous amount of packaging that comes with even the smallest gizmo, the environmental costs of all this electronic stuff are enormous.

Second, and even more important, the constant focus on technological distractions can distance us from our families, our communities, and the world around us. That might sound like the battle cry of an old Luddite hippie, but as we examine in this next series of essays, plenty of evidence shows how too much time spent wrapped up in a virtual world and too little in the real world is bad for us and for our planet. Technology can be an enormously valuable asset, but when it ceases to be a tool for a specific purpose and becomes an end in itself, that's when you know we've lost perspective. And perspective in the digital age is more important than ever.

Without perspective, being constantly *online* and *plugged in* (a phrase meant to evoke the modern computer era, but already outdated) becomes the normal state of being. But being connected electronically is not the same as being connected physically. In fact, paradoxically, being electronically connected all the time has actually made us less social and less community-oriented. Increasingly, we focus on the visual stimuli that captivate us as consumers rather than the gamut of emotions that make us human. We start to see technology as the mother of human inventiveness, not its progeny. We assume that technology will solve all our problems, including the environmental ones—as though there is an Earth version 2.0, just waiting in the wings to be unveiled with great fanfare by Steve Jobs; a slimmer version, perhaps, with more attractive inhabitants and brighter colors.

Yet none of this will solve our problems or make us happy. Ultimately we're not electronic beings; we're biological ones. We have millions of years of evolution programmed into our cells, programming that's infinitely more complex than anything we've created with our shiny new and exciting technologies. Patterns of nature are hardwired into who we are as a species and as individuals. To try to tear ourselves from this biological fabric is not only futile, it's self-destructive. Rather than fighting our biological nature, we need to embrace the fact that most of who and what we are goes back not just a dozen years, but to the beginning of human history and, in some ways, to the beginning of life itself.

*   *   *

BOWING BEFORE THE GOD OF TECHNOLOGY

"According to a new study, air pollution in our city is at its lowest level in thirty years, and we have technology to thank." I heard the words come out of the newscaster's mouth, but I still couldn't believe them. Not the part about air pollution— the part about thanking technology.

In many cities, air pollution certainly has been reduced from levels seen in the early 1970s. Our air is indeed cleaner than it was back then. Of course, with more and more cars on our roads and more and more energy being used, it's starting to get worse again. Still, our cleaner air is a wonderful health and environmental success story—one we don't reflect on often enough or learn from as much as we could.

But we can thank technology? Says who?

Well, it turns out the "study" was a simple analysis done by an industry-funded think tank. The technology angle was theirs and the TV news folks just followed along. On the surface, it's actually true. New technologies and widespread application of existing technologies did help bring air pollution levels down. But no one appears to have asked the simple question, "What spurred the invention and application of these new technologies?" If they had, they would have found the real hero of the story—environmental regulations.

Technology does not arise out of a vacuum. It does not invent itself (at least not yet). It did not wake up one day and decide to clean up our air. Technology is a result of society's values. In the case of air pollution, citizens got angry because their air was dirty and demanded their political leaders do something about it. The result was new environmental regulations that forced industries to clean up their acts.

Of course, many industry leaders did not want to be regulated. Regulation would force them to be innovative, hire new engineers and scientists, fix existing systems, or build new ones. It would cost them money up front. And even if that investment paid off down the road as a result of better efficiency, the initial outlay would cut into their quarterly profits.

So many industries have fought environmental regulations tooth and nail. From car companies to electricity generators and appliance manufacturers, they fought change. They said such regulations would put them out of business. They said conforming to them would destroy the economy. Sometimes they said it couldn't be done—that it was impossible.

But after all the fuss, once the targets and timelines had been set and there was no choice, industries went to work.

They tried new things. They invented new products and processes, and they got the job done. That's where the technology came from. It didn't just appear one day in a burning bush. It was a result of the hard work of a variety of groups of people.

If we recognize that many problems are also opportunities, we can take advantage of them and regulation can help. Our cleaner air is just one example. It's saved countless lives and billions of dollars and improved the lives of millions of people. Phasing out ozone-depleting substances like CFCs is another success. So are seat belts and air bags. None of these advances would have occurred if it weren't for the government regulation that spurred innovation and the people who made it happen.

So, we can thank the engineers and scientists who did the technical work. We can thank the leaders who had the political courage to stand up to nay-saying industrial groups and enact strong regulations. We can thank the health and environmental organizations and the concerned citizens who demanded government take action.

But thanking technology? Sorry, wrong hero.

.   .   .

A SENSE OF WONDER

Humans, I believe, are naturally drawn to lives and worlds outside of our own. We revel in the existence of creatures and even whole societies beyond what we ourselves experience in our everyday lives. But have we gone so far in creating worlds of fantasy that we are missing the joy of other worlds that already exist all around us?

One doesn't have to look far to see examples of the attraction to other worlds in science. From the explorers who first mapped the Earth to researchers trying to understand the inner workings of the human genome and those seeking to find out whether life of some kind exists on Mars, scientists certainly share this sense of wonder. But they hardly hold a patent on it.

Indeed, the trait seems universal. Look at the popularity of fantasy literature or movies like *The Lord of the Rings* and *Star Wars*. Or the escapism of certain video games where other worlds are created for us to explore. This innate sense of curiosity and wonder draws us to each other, to the world around us, and to the world of make-believe.

When I was a child, my escape to another world was a swamp near our house in London, Ontario. It was a wondrous world filled with amazing, bizarre, and beautiful plants, insects, amphibians, birds, and mammals. Every day in that marsh I could always count on finding something new, some exciting new creature or world to discover. Today, that swamp is entombed by a huge parking lot and shopping mall. The vast diversity of life has been replaced by an enormous array of consumer products. What does that mean for youths who spend their time there now?

Eminent Harvard biologist E.O. Wilson has suggested that human beings possess a trait he calls "biophilia"—that is, an innate desire to bond with and understand other life forms. That was certainly true from my own experiences. But I didn't grow up in a world of computers, video games, and the internet. I bonded with my family, friends, and the creatures I found in my swamp. Today's youth, especially in big cities, often lead more isolated, insular lives and can be so far removed from the natural world that they can't even identify the common plants and animals that live around them.

Researchers at the University of Cambridge found that out when they surveyed British schoolchildren. They asked 109 children (boys and girls) to identify creatures depicted on a series of twenty flashcards. Ten cards depicted common British plants and wildlife—things like rabbits, badgers, and oak trees. The other ten cards showed characters from the popular children's trading cards series, television show, and video game Pokémon.

The researchers discovered that at the age of four, children could identify about 30 percent of the wildlife and a handful of Pokémon. But by age eight, children were identifying nearly 80 percent of the Pokémon and barely half of the common wildlife species. They were not even asked to be terribly specific with the wildlife—in many cases answers like "beetle" would have sufficed. Compare that to my father-in-law, who was born in England in 1908 and, as a schoolboy, won a book for identifying some 150 different plant species.

I think this example shows how powerful the need to understand other people, worlds, and life forms really is. When we are deprived of meaningful interaction with the world around us, and sometimes even with our families and friends, we seek to understand and interact with things that exist only in our imaginations or on a computer screen.

Not that the world of make-believe is necessarily bad. The ability to immerse people in a different world through words, images, and sounds is what gives good stories, books, and films their power. And this power is a wonderful thing. The sharing of common stories and experiences can even help us bond with each other as human beings.

But when the world of fantasy, of television, video games, and computers becomes the only outlet for our sense of wonder, then I think we are really missing something. We are missing a connection with the living world with which we share common histories, life cycles, and even segments of our genetic code. Fascinating other worlds exist all around us. But even more interesting is that if we look closely enough, we can see that these worlds are really part of our own.

.   .   .   .

FEEDING OUR SENSES IS IMPORTANT TO HEALTH

We live in a visually oriented world where the vast majority of our attentions are focused on what we can see. Whether we're watching television, working on a computer, or driving a car, sight has become the most dominant sense in modern life. But have we lost touch with our other senses?

Dr. Charles Spence of Oxford University thinks so. He's an experimental psychologist who wrote a report arguing that the use of all our senses is central to our psychological health. According to Dr. Spence, sensory deprivation is common in modern life, and it is harmful to our well-being.

Indeed, most of us have become so accustomed to the dominance of visual stimuli that we don't really think about it anymore. We take it for granted that most of the information we use to understand our world comes through our eyes. It's normal to us. But our other senses may be languishing.

Consider smell. Smell is one of our most powerful senses. It's directly connected to the part of our brain that processes memories and emotions. Yet most of us live and work in largely sterile, odor-neutral buildings. Most of the odors we do smell indoors are overly perfumed commercial items like soaps and air fresheners. Outdoors, walking on our city streets, we shut out the noise and stink of automobiles so we can focus on where we're going.

And what about our sense of touch? According to Dr. Spence, children may be growing up "touch-hungry" because people do not touch each other often enough and are not getting enough tactile sensations. He argues that we should be putting more tactile objects into our schools and workplaces to help stimulate this sense.

Dr. Spence may be tapping into a problem that is deeper than our five senses. Modern life in the industrialized world is often far removed from natural rhythms that, for most of human history, have played a major role in our existence. For example, now that we spend 90 percent of our lives indoors, we are no longer as attuned to the change of seasons. We keep our climate-controlled homes at warm spring temperatures. "Summer" fruits and vegetables are available all year round, as is "farmed" salmon. Modern offices often offer little natural lighting to even indicate the time of day!

The relevance of day and night to modern society has changed. Grocery stores, fast-food outlets, and even gyms are open twenty-four hours. And have you ever been to one of those fancy video arcades with interactive games? I went to one with my grandson. I confess: I had a good time. They're fantastically loud, hyperkinetic places that overwhelm the senses—especially our vision. And nowhere—nowhere—will you find a clock.

It's intentional. It's also no accident that there are no windows in these modern-day cathedrals of technology. Nor is it coincidental that the sounds and lights are so hypnotic and mesmerizing. The owners want you to lose track of time. They want you to spend hours plugging the machines with tokens. They want you to be completely disconnected from reality. This is true for casinos as well. No clocks, no windows. Just the ringing of slot machines and the clatter (through a digital sound effect) of cold, hard cash.

Ironically, some of the most popular video games are those that emulate real-life experiences. You can ride a mountain bike, skateboard, or snowboard, or drive a car, ski—even fish! But again, 90 percent of these experiences are visual only. You don't smell the salt air of the ocean or feel the swell of the waves under your boat or the slipperiness of a freshly caught fish.

It is disconcerting that even in our spare time we flock to malls and arcades for virtual experiences rather than the real thing. There's nothing wrong with a little visual stimulation, but there's more to life. We mustn't forget to feed our other senses. Right now, they're starving, and that may not be good for our mental or physical health.

.   .   .

HARNESSING THE POWER OF STUFF

Modern life is a communications paradox. We are in touch with each other as never before—cell phones, internet, text messaging, email. In fact, we now have to make an effort *not* to be in touch with other people.

Yet, at the same time, it can be harder than ever to reach large groups of people. Television audiences are fractured. Newspaper readership is down. Magazines drift in and out with the change of seasons. Town hall meetings go unattended.

So what binds us together? What's the cultural glue that we all share—those common touchpoints we all understand? Unfortunately, it seems the answer today is—stuff. The power of globalization means that most of us are buying the same products, wearing the same clothes, eating the same food, and shopping at the same stores as our neighbors. Brand logos are now among the first things children recognize.

Because I am a well-known Canadian, I frequently get approached to promote this same stuff, endorse products, put my name on things, and generally sell out in every conceivable fashion. Do kids still use that term—"sell out"? I don't know if they do. It's hard to imagine they would now that every movie star, pop icon, and über athlete seems to shill at least one product. Sometimes dozens.

Generally, I ignore such requests. But is that always wise? The fact of the matter is that today, stuff-selling mega-corporations have a huge influence on our daily lives. And because of the competitive nature of our global economy, these corporations are generally only concerned with one thing—the bottom line. That is, maximizing profit, regardless of the social or environmental costs.

Which brings us back to our paradox. If we want to move towards a low-polluting, sustainable society, we need to get consumers to think about their purchases. But how the heck can you reach them when methods of communicating are so fractured? Well, what if we could harness the power of stuff and turn it against itself? A few years ago, I was asked by a coffee company to put a quote on their disposable cups to "spur coffeehouse-type discussion." It could say whatever I wanted—perhaps even question the product itself. The list of other well-known individuals whose quotes would grace the sides of cups was long and distinguished. I would have been in good company.

Now, make no mistake about it. The only reason the company was doing this, really, was for brand promotion. Having the words of well-known people on the cups would make the company look good. Customers would, consciously or unconsciously, associate the person with the brand. If the person were generally held in high esteem, the brand would benefit.

But here was also an opportunity to get to people right at their cultural touch point—stuff. Here was a chance to ask them to think about the choices they make every day and how those choices affect our future. And it was an opportunity to do it right when they were making the decision. Right when their choice really mattered. Here was a chance to use the power of a giant corporate machine to get people thinking.

And I balked. I couldn't do it. Time and time again, people I asked about the proposition said, "Are you nuts? That will look like an endorsement, no matter what you say." Sadly, I think they are right. Most people simply wouldn't be able to get beyond the medium to the message. They would assume I was tacitly endorsing the product—or worse, getting paid for it.

Maybe people aren't quite ready to accept the idea of questioning a product on the product itself. Perhaps it's a bit too postmodern even in this day and age. But with today's fractured communications world, unless we look at new ways to reach people and get them to question their choices, most will choose what is cheapest or most convenient—choices that have essentially already been made for them by corporations with their eye on the bottom line.

For the sake of our future, this is one paradox we had better solve.

BRANDING KIDS STARTS EARLIER THAN EVER

Would you let your kids play in a swamp? Odds are that most parents would balk at such a notion today. After all, a swamp seems so dirty and teeming with who-knows-what. But if not a swamp, what about a forest or a creek—or even a backyard? What worlds are children exploring today, and what are they learning from them?

As I have mentioned, my childhood playground was a swamp near my home in southern Ontario. I spent countless hours there, catching tadpoles and wading though cattails, delighting at each new discovery. As a result, I could easily name dozens of species of birds, fish, and insects. This was my world, and it shaped who I am today.

But while my world was full of nature's delights, today's children face a world dominated by consumer delights. Instead of a real swamp, their world is often "virtual," consisting more of television, video games, and the internet. Each of these technologies wields tremendous power, and children can learn a great deal with them. What they learn, however, is not necessarily what we intend.

Advertising certainly existed when I was growing up, but it was nothing close to the saturation levels faced by children today. In my swamp, there were no billboards. Frogs did not croak "Coke." Birds did not pull banner ads. The swamp was not sponsored by an oil company. And I was blissfully free of the consumer messages that bombard children in the twenty-first century.

So while I learned the names of other living creatures, kids today are far more likely to learn the names of various products and popular brands. And according to new research, this constant assault of brand imagery is reaching our children at earlier and earlier ages.

A study published in the *Journal of Applied Developmental Psychology* has found that children as young as two are now able to recognize common brand names. Researchers tested some two hundred Dutch children, presenting them with common logos, such as McDonald's, Nike, Mercedes, and Cheetos. Most two- to three-year-olds recognized eight out of twelve logos, and the majority of eight-year-olds recognized 100 percent of them—including Camel cigarettes and Heineken beer.

Researchers found that one of the strongest correlations with higher brand recognition scores was a child's exposure to television. Generally, the more television a child watched, the more readily he or she was able to recognize brands. This makes sense, given television's power as a visual medium.

But the researchers also point out that their results should be a warning about the potential for advertising to influence the most impressionable minds. Advertising to infants and toddlers is a rapidly growing trend. Just ten years ago, most marketers targeted children only over age six. Today, with the success of toddler-based television shows like *Teletubbies,* researchers say infants and toddlers have been identified as a "vital and undeniable target group."

In fact, the authors argue that marketers have already done their own research about the cognitive and behavioral effects of advertising on young children. In most cases, however, the results have not been made available to academics or policy-makers. In other words, marketers aren't just incidentally targeting some of the most vulnerable members of society—they are actively targeting them and then keeping quiet about it.

Children of the twenty-first century are growing up in a world very different from the one I faced. In some ways they have more opportunities than my generation ever did. But they also face new problems and threats that we never would have imagined. Given the insidious nature of some of those threats, maybe a swamp isn't such a bad place to play after all.

.   .   .

REALITY TV THE CLOSEST
SOME CHILDREN GET TO REALITY

In the heat of the summer, do you know where your kids are? According to a 2006 study, they're probably in a darkened room somewhere, staring at a television or computer monitor.

The study, published in the *Journal of Environmental Management,* found that per-capita visits to U.S. national parks have been declining for nearly twenty years—largely as a result of people's increased time spent watching television and movies, playing video games, and surfing the web.

Although the study was conducted in the U.S., and Canadians tend to have stronger ties to the outdoors, I would be surprised if the trends were that different in Canada. Canadians watch less television than do Americans, but they also have some of the highest internet usage rates in the world. They stare at computer screens more than practically anybody else.

And while lower attendance levels in national parks do not necessarily mean people are spending less time outdoors in general, the connection to time glued to electronic media is hard to ignore. In fact, the evidence was strong enough for the researchers to conclude, "We may be seeing evidence of a fundamental shift away from people's appreciation of nature (biophilia, Wilson 1984) to 'videophilia,' which we here define as 'the new human tendency to focus on sedentary activities involving electronic media.' Such a shift would not bode well for the future of biodiversity conservation."

Indeed. The internet is a fantastic tool, as is television. Even video games can have educational value as well as be entertaining. But as with anything, there needs to be a balance. When I was a boy, escaping to the air-conditioned comfort of a movie theater during the heat of the summer was a real treat. But it was an exception, not the norm. Far from spending the majority of my time indoors, I spent most of my waking hours outside— swimming, fishing, hiking, or just exploring.

Times certainly change, but when our behaviors change in a way that alienates us from the natural world upon which we depend for our food, our energy, our natural resources—our very lives—that, to me, is cause for concern.

We tend to forget that the world we live in today—the electronic age—barely registers in the timeline of human history.

For the vast majority of that history, we were a rural people. We lived in family groups and small villages and followed the natural cycles of days and nights and the seasons. We didn't buy processed food from the mini-mart, text-message people halfway around the world, or watch infomercials at 3:00 AM bathed in the glow of artificial light. Most of the modern electronics we take for granted today have been around for only fifty years or less.

These electronics may make our lives easier, but I often question whether they are making our lives better. People tap away on BlackBerries and personal computers during meetings. They take cell phone calls during the birth of their children and play video games for days at a time, virtually without a break. They walk down the street listening to MP3 players, lost in their own worlds. We seem to be plugged in twenty-four hours a day, seven days a week. That strikes me as decidedly unbalanced.

So try this: for a month next summer, or maybe just a week, or even a day—unplug. Put away all your electronic gizmos and go outside. Lie under a tree. Watch the clouds. Smell the air. Enjoy real life, rather than a virtual version of it.

Most important, take the kids.

.   .   .

THERE OUGHT TO BE A LAW

Sometimes I feel like I'm the only person in Canada who doesn't own a cell phone, and I don't think I ever will. Watching people barking into their phones at the gym, on ski hills, and in restaurants, I wonder why they bother to go there in the first place. But that's their personal choice. What really bugs me is the planned obsolescence of so many of these technologies.

Sustainability is a word generously slathered through corporate and government reports. It flows freely from the lips of those who say they're committed to being green, but to turn words into action we've got to reflect on what those words mean in everyday life. Sustainability is about ensuring that what we do today does not compromise the opportunity and future for our children and grandchildren. We are a long way from achieving that right now. Everything we use—food, clothing, energy, consumer goods—everything comes from the biosphere, the zone of air, water, and land where life exists. And all of our garbage, effluent, and waste goes back into that same zone of life. Indeed, if the globe were reduced to the size of a basketball, the biosphere would be thinner than a layer of plastic sandwich wrap. That's it, our home where we live. It is finite and fixed; it can't grow.

Today most people, myself included, are all agog at the wondrous outpouring of new technology—cell phones, iPods, iPhones, laptops, BlackBerries, and on and on. Even though I am techno-incompetent and like to think I shun these new devices, I actually have a drawer filled with the detritus of yesterday's hottest products, now reduced to the status of fossils. I have video cameras that use tapes no longer available, laptops with programs incompatible with anything on today's market, Beta cassette recorders, portable tape and CD players I no longer use, and more. But what really upsets me is opening a drawer and finding it filled with cords, chargers, and transformers for which there is no longer anything to plug into. Yet if I misplace a cord to charge the battery of my current camera or laptop, none of the cords in the drawer works!

Forgive my rant, but not long ago, I embarked on an epic search for a cord to plug into my wife's cell phone to recharge it. We were in Toronto and the poor phone kept bleating that it was running low and the battery needed recharging. Calls were coming in to Tara but there wasn't enough juice to return them. We asked others in our group to lend us a charger but found every single one was incompatible with her phone.

So we began a search—from big-box technology super-stores to smaller suppliers and the cell phone companies themselves—all to no avail. Finally, a salesperson told my wife, "That's an old model, so we don't stock the charger any longer."

"But I only bought it last year," sputtered Tara.

"Yeah, like I said, that's an old model," he replied without a hint of irony or sympathy. So in the world of insanely rapid obsolescence, not only does each company's products have its own unique plugs and cords, each successive model is incompatible with the previous one it replaces.

If there must be new models with new gimmicks every few months, why can't there be a single charger or transformer that can be used interchangeably by all companies' products and from year to year? Why can't there be some sort of standard? How technologically advanced is a cord that it must be replaced with a new model every six months? The proliferation and sheer waste of this type of practice is mind-boggling.

Someone has to pay for all those disposable cords, chargers, and adapters, to say nothing of the products themselves. That someone is all of us. And not just for the product, but also for the pollution created when it's made and disposed of—right back into the biosphere. It's time for producers to take responsibility for their products' entire life cycles and not just pretend they can wash their hands of the problem when it goes out the door.

·   ·   ·

GET OUTSIDE—IT'S GOOD FOR YOU

When your mom told you to go outside and play, it seems that she really did know what was best for you. Just being outdoors or having access to the natural world has been proven to have physical and mental health benefits. And research has now found that the more diverse and vibrant an ecosystem is, the healthier it is for us.

One of my personal favorite places in the world is Haida Gwaii—the Queen Charlotte Islands—off the coast of British Columbia, Canada. The diversity of life there in the cold, nutrient-rich waters, on the shorelines, and in the old-growth forests is simply astonishing. I'm hardly alone. Lodges and retreats are popping up all along B.C.'s pristine middle and north coast as people search for places to get away from the stress of their everyday lives.

People gravitate to these kinds of places, they usually say, because they are beautiful, peaceful, or relaxing. Sometimes they will venture as far as describing experiences with these ecosystems as uplifting, moving—even spiritual. For others, it's a feeling that's difficult to describe in words, but being in nature just somehow makes them feel better.

Although many people may not realize it, there's actual biological value in having experiences with nature, value that is measurable and quantifiable. It's long been established that general health, mental fatigue, and physical injury all recover faster when patients have access to natural areas. Studies have shown, for example, that surgery patients recover more quickly when they have views of natural landscapes outside their windows rather than views of bricks and concrete.

Some people attribute this connection with nature to the perceived benefits of having access to fresh air and fewer distractions. But it actually goes much deeper. Famed Harvard ecologist E.O. Wilson calls this connection to the natural world biophilia. It's a term he coined and it simply means that he believes humans have a kinship with other living things that is genetically programmed into us.

So I'm sure Dr. Wilson wasn't the least bit surprised by a 2007 study published in the science journal *Biology Letters*. The study found that the psychological benefits of urban green spaces increase with the diversity of life found in them. Researchers interviewed more than three hundred park-goers in the medium-sized city of Sheffield, England, and compared their answers to an analysis of the species richness, or biodiversity, of their parks.

They found that while the overall size of a park influenced the visitor's perception of how it made them feel, even more important was the diversity of life. Bigger parks made people feel better, yes. But species-rich parks were even more beneficial. In fact, the researchers report that visitors to the green spaces were actually able to consciously perceive differences in species diversity—especially with plants.

As it turns out, when it comes to our health and well-being, not all parks are created equal. A grass field, for example, is likely to be far less beneficial than a natural area with a greater diversity of plant and animal life. We now know that humans, consciously or otherwise, are able to judge the overall diversity and vibrancy of green spaces. What's more, the more diverse and vibrant those ecosystems are, the greater their value to humanity in terms of our own personal health and well-being.

With three-quarters of North Americans now living in urban areas, citizens must ensure that city planners and municipal politicians are paying attention to this kind of research. It underscores the need both to protect our most diverse ecosystems and to design cities to have more and larger green spaces. Ultimately, our health depends on it.

RESPECT FOR NATURE HAS TO START AT HOME

One of the refrains I often hear from people is about how we have to educate our kids to be more environmentally responsible. It's too late for us, they say. Adults are too set in their ways to change. We've got to teach the children!

What a cop-out. So not only are we leaving our children with a legacy of global warming and other environmental challenges, but we'll leave it up to them to fix the problems we created? Sorry, but that's fundamentally unfair. It also sets us all up for failure. Children do what we do, not what we say. If we don't change our ways first, what incentive do children have to behave more responsibly?

This isn't to say that we couldn't be doing a better job with teaching children in school to be more environmentally aware. But doing that isn't just about pointing to things in a book. It's about doing things differently together. It's about changing behaviors.

Even before kids get to the classroom, look at how they get there. Every year it seems a group releases new, alarming statistics about childhood obesity. More and more, we hear about how kids are becoming increasingly housebound and sedentary. Yet we fail to connect these problems with the lineups of SUVs and minivans several blocks long outside schools every morning and every afternoon.

Chauffeuring our children to and fro not only denies them an opportunity to exercise, pollutes our air, and adds to global warming, it further removes them from the natural world. We don't respect things that we don't understand. And it's very hard to understand something without experiencing it.

At the risk of sounding very old-fashioned and very old, when I was young I walked to school. When I got older, I rode a bicycle. So did everyone else. It's a great way to get exercise and experience the outdoors. There's nothing like walking to help you get to know your community. And not just the people, but the plants and the trees, the animals, the weather, and the seasons.

Reconnecting our children with nature in their everyday lives is the first step in an environmental education. That means getting children outside into the world to experience it firsthand rather than through TVs, computers, or YouTube.

I'm not saying that there isn't a place for technology in helping us understand the world. After all, I've spent forty years trying to educate people through television. In fact, one of my favorite tools is an addition to Google Earth called Google Sky. For years, I've been a fan of Google Earth as a tool helping people understand just how small our world really is and how connected we all are. Google Sky adds a new dimension, as now you can turn the lens around and look at what's out there. It's really an interactive chart of the heavens, the stars, and the planets that lets us explore the universe and ultimately better understand our place in it.

But as fascinating as it is, nothing can replace the real experiences we have outdoors, peering through a telescope into the night sky. Or digging in a garden. Or exploring a swamp, a forest, or a tide pool. If we want our children to be more environmentally responsible, we have to show them why they should be. We have to emulate the right behaviors and teach them why environmental sustainability is so important.

So yes, this means we need more schoolyard gardens, better outdoor education curricula, more field trips, and more sustainable schools. But it also means we need more exercise. More cycling and more plain old walking. We have to get our kids outside more to play and explore the wonders of nature so that they will come to understand it better.

This isn't just up to kids or teachers. It's up to parents. It's up to school boards. It's up to all of us to ensure that we're not telling our children one thing and doing another. Anything else and we're not just lying to them. We're lying to ourselves.

.   .   .

CHRISTMAS COMPLAINTS MISS THE POINT

It can start in October—even before Halloween. The television commercials, the flyers in the mail, the decorations in the mall. Christmas is now a two-month event—one long blowout sale.

But there's also no shortage of people decrying the commercialization of the holidays. The criticism itself is nothing new. People have been complaining about it for decades. Every year, the Christmas season gets a little longer, and every year people complain about it a little more.

It's certainly a valid criticism, one that I can't help but make myself. As the holiday hype escalates, so too does our accumulated waste. The roads become packed with anxious shoppers driving from mall to mall in search of the right gifts. The malls become stuffed with Christmas goods and trinkets, all vying to catch the shopper's eye. And the shoppers themselves become stuffed with holiday sweets and extra-large gingerbread lattes. The whole enterprise is a monument to excess.

For some, this excess typifies everything that is wrong with the developed world. We consume far more than our share of the world's resources. We create huge amounts of waste. We obsess with fads and fancy while species die out, pollutants seep into the food chain, and the climate changes. Christmas is the pinnacle of our hyperconsumptive lifestyles, so it's easy to point a finger and condemn the whole stressful, chaotic, overindulgent experience.

But the real question is why? Why do people put themselves through all the stress and pressure? Why do we go into debt so we can give gifts that the receiver probably doesn't even need? Two months after Christmas, how many of those gifts can we even remember? And why do we complain about the excesses of Christmas and then fall for it again every year?

I believe we are trying to fill a void. With fewer and fewer people taking part in the religious aspects of the holidays, many are looking for other rituals to take their place. Humans have an innate need to connect with their families, their communities, and the rhythms and cycles of nature. Throughout human history, we've done that with celebrations and rituals to mark the changing seasons, the lunar cycles, and important stages in our lives.

But today's world is very different, and in many ways, it runs against millennia of the human experience. This new world runs 24/7. This world is built on consistency and uniformity rather than reflecting natural rhythms and local cultural or geographic differences. This world has few rituals to reflect the stages of our lives, the changing of the seasons, and the passage of time. It doesn't matter if it's dark outside. We just turn on a light. It doesn't matter if it's cold outside. We just turn up the heat. The seasons may change, but our work schedules stay the same. Fresh vegetables and fruits are available year-round regardless of whether or not they are in season or grown anywhere nearby. A Big Mac is a Big Mac, here or in Turkey.

This world we've created is hard on the planet, and it's hard on us. We've tried to isolate the human experience from the rest of nature, but it's an impossible task. Humans are a part of nature. Whether we like it or not, our bodies respond to changes in the natural world. The more we try to deny who we are, the less connected we will feel and the more damage we will do to the planet.

In the absence of God or spirituality, in the absence of a capacity to respond to seasonal patterns and natural rhythms, and in the absence of meaningful social rituals, people are grasping on to whatever they can to help ground them in their communities. If that means spending days at a time in a crowded mall, then that's what we do. That becomes the ritual. That becomes Christmas.

I think people are hungry for change but feel trapped. We are yearning for meaning but accepting baubles and trinkets instead. Until we stop denying our biological roots and embrace our humanity, we will never find the meaning we seek. It's just not something you can pick up at the mall.

9

LIGHTS, CAMERA,

SOUND BITE

Social change and the media

GLOBAL WARMING IS GENERALLY considered to be one of the biggest challenges humanity will face this century. Lead scientists and economists all say it could take an enormous toll on our environment, our health, and our economy. Naturally, this has people concerned. Google "global warming" and you get quite a few hits—about forty-two million, in fact.

Now Google "Britney Spears": seventy-eight million hits.

Humanity, we have a problem.

True, these are just web pages, and any deranged fan could put up hundreds of pages devoted to his favorite singer. Surely our hallowed news media do better. Our proud fifth estate must have more important things to do than follow an apparently mentally ill pop star around. Indeed, according to Google's news index, which looks at media stories, global warming did garner 98,000 news headlines in 2007, compared to 34,800 headlines for Britney. You heard it here first, folks. According to the media, global warming—a problem that threatens to cripple our environment and our economy, or at the very least radically change the way we live—is officially somewhat more newsworthy than Britney Spears.

How did we get to this point as a society? More important, how can we get beyond it? Social change is never easy. If you look at great social advances in the past century, from universal suffrage to civil rights, it always took enormous efforts to change society, and the media—that bastion of the free dissemination of information—played a key role. But the world has changed a lot in the past twenty years. News used to be serious and deal with actual issues. Today, news is more focused on entertaining, rather than informing. It looks for cheap scandal, conflict, and titillation. It reduces complex stories to black and white. It neglects to probe and question. And in the process, it fails us as a tool, as a service, and as an institution.

I have an enormous respect for journalists. But today my respect is tempered with pity and sadness. In today's fractured media world, can newspapers afford to be serious and face ever-decreasing readership? Can a journalist do her job when she's faced with few resources, tight deadlines, and an inability to follow up on important stories because they are no longer "new"? In such a world, where can we have a good discussion about important issues?

In this series of essays we look at how the media portray environmental issues and the effect that can have on the public mindset. That effect can be quite profound. We lost nearly two decades in the fight against global warming because the media were either complacent or complicit in giving a handful of climate-change skeptics the same amount of airtime and the same number of column inches as the vast majority of legitimate climate researchers who had the vast majority of evidence on their side. Somehow, this kind of portrayal of issues has come to mean "balance" in journalism. As a result, for years the public thought that there was a debate among scientists about the causes of global warming and even whether it was actually happening. This left people concerned, but confused—willing to sit back and wait until the "debate" was over, even though for more than 95 percent of climate researchers the debate had long been settled. Without any serious public pressure, politicians saw no need to press the issue, and without political will, nothing happened.

Conversely, too many negative headlines about environmental issues can also cause a problem. Rather than empowering people to seek change, it can lead them into despair, to a "bunker" mentality where, instead of working with their neighbors to change the world, they actively seek to disengage from community life, becoming more isolated and reclusive. If headlines are all about how the world is falling apart, how could one person possibly make a difference? Up against such great odds, what's the point of even trying? After all, no one wants to worry about environmental problems like global warming. They're on a scale that dwarfs the individual person. In such a situation, and without ready solutions, it's easy to see why people would choose instead to focus on things they do have control over— their mortgages, their cars, their TVs.

Yet despite all this, in the past two years environmental awareness has blossomed. The latent public concern that for so long was buried under bad news or soothed by climate-change skeptics has come to the forefront again. Hybrid cars have become status symbols among the environmentally aware. Bringing reusable bags to the grocery store is rapidly becoming the norm rather than an eccentricity. All of this is good news. Sometimes these small, symbolic gestures can be the first step towards a profound change in the way we see the world. Other times, the commitment ends at small gestures. Which one will this be? As television news reporters love to portentously say, only time will tell. A great cliché to cap the evening news, sure, but one that also conveniently ignores the fact that the media are more than just a reflection of society; they are active participants too.

.   .   .

THINKING CRITICALLY ABOUT INFORMATION SOURCES

Who do you turn to for news and information about science and health issues? Television? Newspapers? Industry groups? The internet? Recent polls have found that public trust in many of these sources is very low. The most trusted sources of science and health information tend to be experts and medical professionals as well as non-government organizations. At the bottom end of the trust spectrum are industry groups.

The public has good reason to be skeptical. Health and science issues are particularly open to manipulation, since they are often complicated and hard for the layperson to grasp in short articles or stories. And it seems that some groups will go to any lengths to distort scientific information. For example, a 247-page report released by the World Health Organization (WHO) in 2000 details how tobacco companies have systematically sought to undermine WHO's research and efforts to curb tobacco use for more than a decade.

Tactics used by the tobacco companies included paying WHO employees to spread misinformation, hiring institutions and individuals to discredit the WHO, secretly funding reports designed to distort scientific studies, and even covertly monitoring WHO meetings and conferences. Confidential tobacco company documents do not mince words as to the industry's goals, using phrases like: "Discredit key individuals," "Attack

WHO," and "Work with journalists to question WHO priorities, budget, role in social engineering etc."

The report also notes that, "Tobacco companies have conducted an ongoing global campaign to convince developing and tobacco-producing countries to resist WHO tobacco control policies." These tactics appear to have been successful, as is evident from the growth in cigarette sales in developing nations like China. There are now more than 300 million smokers in China consuming 1.7 trillion cigarettes every year—figures that have tripled since 1978. In the coming decades, smoking is expected to surpass infectious disease as the leading threat to human health worldwide.

Profit is the obvious motive for the tobacco industry's manipulations, but the media are not immune to distortions and fabrications of their own. For example, in a segment on organic food appearing on ABC's vaunted 20/20 news program several years ago, a reporter actually fabricated some scientific test results and distorted others to make his story more sensational and fit the desired angle.

In the segment, the reporter said that tests by ABC found no pesticide residue on conventional produce, so buying organic produce to reduce exposure to pesticides was essentially a waste of time. After the segment aired, the U.S.-based Environmental Working Group questioned the claim and found that no such tests were ever actually conducted. In fact, tests by the U.S. Environmental Protection Agency have found residue from at least twelve different pesticides on 90 percent of conventionally grown celery, 53 percent of lettuce, and 26 percent of broccoli.

The ABC reporter also misrepresented tests for *E. coli* bacteria to make organic produce actually seem dangerous. At one point, he held up a bag of organic lettuce and confronted the head of the organic industry's trade association, proclaiming: "Shouldn't we do a warning that says this stuff could kill you and buying organic could kill you?" With no evidence to back up these claims, the reporter has since apologized and been reprimanded by the network.

Even peer-reviewed scientific journals can sometimes suffer from distorted or even fraudulent scientific claims. In an editorial in *Science,* for instance, magazine staff reported that the journal had unknowingly published a study based on manipulated data, thus invalidating not only that study, but another that had been based on the original article. Such cases are rare, but they can happen.

There is no way to be 100 percent certain of the veracity of any science news, but we can use our best judgment to achieve reasonable certainty. We have to weigh the balance of evidence, consider the credentials and motives of the source, and think critically. It's not always easy, but that's the price of being truly informed.

.　　.　　.

CLIMATE PAPER TWISTED BY CONTRARIANS

Climate science isn't easy for most of us to understand at the best of times, but it became even more confusing in 2000 after a climate expert released a controversial paper on the topic.

In his paper, James Hansen of NASA's Goddard Institute for Space Studies pointed out that while many researchers have focused largely on the effects of rising carbon dioxide emissions on our climate, other greenhouse gases are equally important and may actually be easier and cheaper to reduce in the short term.

After the article was published, some newspaper editorials declared that Dr. Hansen had changed his mind about global warming and was suggesting that we should scrap the Kyoto Protocol, the international agreement to reduce greenhouse gas emissions. It was interesting to watch as pundits twisted Dr. Hansen's article to fit a predetermined agenda but disturbing to think that many people believed them.

One opinion piece that ran in several newspapers across Canada stated, "there can be little doubt that dropping Kyoto's carbon dioxide program is exactly what Mr. Hansen is proposing." Really? Why? First, Dr. Hansen did not say we should drop Kyoto. In fact, he said of his scenario: "This interpretation does not alter the desirability of limiting CO_2 emissions."

But more importantly, the Kyoto Protocol does not single out carbon dioxide from other greenhouse gases for reduction. That wouldn't make any sense. Greenhouse gases all have a similar effect—they trap heat, and that's what's causing our climate to change. Some gases, like methane and chlorofluorocarbons (CFCs), are much more powerful than carbon dioxide. The goal of the Kyoto Protocol is to slow climate change, and that's why it deals with six different gases, not just one.

What Dr. Hansen said was that reducing some of those other gases first might be cheaper and more effective than going after carbon dioxide right away. That's an interesting theory, but it was also controversial—something some editorials also neglected to mention. One editorial actually stated, "it is impossible to deny his conclusions," as though Dr. Hansen's paper would be the last word ever written on the subject.

Hardly. Many scientists questioned some of the assumptions in Dr. Hansen's paper, such as the expected growth rate of carbon dioxide emissions. A subsequent article in *Nature* quoted Harvard University professor John Holdren as saying, "it's all too easy to get the impression from the article that CO_2 is not as important as had been thought, and that is not correct."

By using Dr. Hansen's paper to advance their "do nothing" approach to global warming, the climate-change skeptics may have actually shot themselves in the foot. Dr. Hansen was hardly a shrinking wallflower of a scientist who would stay quiet and lay low while others misused his research. In fact, Dr. Hansen has turned out to be one of America's most outspoken advocates for reducing greenhouse gas emissions and one of the most vociferous critics of the Bush administration. In addition to his scientific research, he has been a passionate advocate for change to prevent dangerous levels of global warming.

Of course, those using his paper in 2000 as a springboard to "disprove" climate change or to undermine international efforts to slow it did not know that Dr. Hansen would be such a bulldog on this issue—even taking the Bush administration to task for misusing science and altering his written testimony to suit a political agenda. But even in his original paper, Dr. Hansen went to great lengths to offer solutions to help slow warming, like switching from coal-fired to gas-fired electricity-generating stations and changing government policies to encourage energy efficiency and the use of renewable energy.

Dr. Hansen had not changed his mind about the need to reduce greenhouse gas emissions, and his ideas could have easily been implemented through the Kyoto Protocol. So why twist his conclusions? Why omit and ignore key sections of his paper? Dr. Hansen may have stirred up some debate with his paper, but the real confusion was concocted in the media.

.   .   .

WHO ARE THE REAL SURVIVORS?

You'd have to have been living under a rock to never have heard about the mega-hit television show *Survivor*. Somehow this "reality" game show actually became news and regularly made its way to the front pages of newspapers across North America.

The first game revolved around sixteen contestants stranded on a desert island in the South China Sea who competed to "Outwit, Outplay, and Outlast" the others for thirty-nine days to win the prize of US$1 million.

It was a novel idea, to be sure, and viewers seemed to enjoy the show's social dynamics and exotic tropical setting. But unintentionally the show also gave us an indication of how far removed people in the developed world are from the reality of life that still faces the majority of the inhabitants of our planet.

Human civilization has existed for thousands of years. But most of the modern conveniences that the contestants so badly missed, like telephones, refrigeration, and television, have only been around since the turn of the last century. Some, like iPods and email, are less than a decade old!

Like most people in the developed world, the majority of the castaways on *Survivor* had a hard time fending for themselves without the help of modern technology. They managed to crack a few coconuts and find some tapioca, but other than that they had no idea which plants were edible and which were poisonous or how to obtain enough protein. Only one member of the group (the eventual winner) was capable of catching fish.

Although the contestants were certainly roughing it, they were hardly abandoned. They received ample rice to eat and had access to basic medical items like insect repellent, sunscreen, Band-Aids, and iodine. And they had helicopters waiting to airlift them to a hospital in case of an emergency.

But seemingly lost on the contestants, crew, and producers was the fact that, while the survivors pined for "normal" food like Big Macs and pizza, not too far away, millions of people actually were living off a couple of bowls of rice a day—many for their entire lives. These people are not even afforded the very basics of health care, like vaccines or antibiotics. For them, insect repellent to help ward off malarial mosquitoes is an unattainable luxury.

According to the World Bank, the real survivors are the 1.3 billion people who live on a dollar or less a day and the 3 billion who live on two dollars or less a day. While the contestants and crew of *Survivor* complained about the lack of hot showers and gourmet food, 1.4 billion people still do not have access to safe drinking water.

In Indonesia and Vietnam, near the island where the first *Survivor* season was filmed, more than one-third of children are underweight. In some South Asian countries, up to 50 percent of children are born with low birth weights caused by malnutrition.

The reality is that *Survivor* is a rich person's game. We can afford to be titillated by the idea of struggling to survive with a bare minimum of resources. For us, it's a spectator sport—a pleasant diversion from our hectic modern lives. For the contestants, it was fun because they knew they got to go home at the end and one would be $1 million richer. Their Asian neighbors have no such incentives.

If the lesson learned by contestants on the show was that they should not take modern life for granted, perhaps the next *Survivor* should take place in the slums of Calcutta or Manila. Then perhaps contestants wouldn't just learn to appreciate all that they have at home, but also what most others do not.

⚬ ⚬ ⚬

THE GOOD, THE BAD, AND THE UGLY OF SKEPTICISM

Skeptics are a strange bunch. Certainly, a healthy dose of skepticism is essential to science and crucial to helping us all weed out truth from the everyday noise of media and advertising. But some, despite overwhelming evidence, still cling to theories and beliefs that have little basis in fact. Some of these "skeptics" deny evolution, others deny that the world is round, and, increasingly, some deny that human impact on the environment poses a serious threat to society and to life in general.

Intuitively, that just doesn't fly. Every day, it seems, there's another troubling news report about pollution, global warming, or species extinction. It's depressing. Actually, some analysts say that the "eco-guilt" brought on by all these news stories is what's fueling the popularity of skeptics as people search for more comforting worldviews.

Skeptics often label people (like me) who disagree with them as "doomsayers." Skeptics are tired of doomsayers, who they say espouse nothing but bad news. And they offer an enticing alternative: everything you hear about the sorry state of our environment is wrong. The doomsayers are wrong; the scientists are wrong; the news stories are wrong. The environment is actually improving.

Perhaps that uplifting conclusion explains why a new book by Bjørn Lomborg received so much publicity when it arrived in North America. That book, *The Skeptical Environmentalist,* claims to measure the "real state of the world," which, according to Mr. Lomborg, is just fine.

Of course, another possible reason the book received so much publicity is because of the deep pockets and influence of some big businesses that have vested interests in maintaining the status quo. Indeed, Mr. Lomborg's take on the state of the planet is very similar to the positions of some large industry-funded institutes and groups, such as the now-defunct Global Climate Coalition. These groups wage big-budget campaigns to confuse the public about issues like air pollution and global warming.

So it was strange to see Mr. Lomborg's book and his views go largely unchallenged in the media, which often completely ignored the fact that his beliefs ran contrary to the vast majority of scientific opinion. In fact, before the release of his book in North America, Mr. Lomborg's views had already been widely discredited by many of his colleagues at the University of Aarhus in Denmark. Some had even written articles with titles like "Lomborg's claims are untrue and dangerous" and "Lomborg's facts are absurd and irrelevant."

Yet, despite this criticism from his peers, one of Canada's national newspapers ran an unchallenged series of full-page essays from Mr. Lomborg as well as numerous editorials and a book review so over-the-top in terms of praise that it was almost comical. The reviewer even had the audacity to conclude, "After Lomborg, the environmental movement will begin to wither."

Such outrageous claims should have immediately raised suspicion. A much different review of the book appeared in *Nature,* one of the world's most respected science journals. It described *The Skeptical Environmentalist* as "a mass of poorly digested material, deeply flawed in its selection of examples and analysis." The review went on to say that Mr. Lomborg's "bias towards non-peer-reviewed material over internationally reputable journals is sometimes incredible . . . At other times it seems fictional."

A critical eye is a valuable tool, and the media are supposed to help us in this regard by presenting information in a balanced way so the reader, viewer, or listener can make an informed decision. Unfortunately, balance is often lost in the desire to be controversial or shocking or to meet a desired editorial slant. That's a shame, because issues that pertain to the health of the planet are far too important to be treated as mere journalistic fodder to appease the skeptics.

THE SKY IS FALLING! THE SKY IS FALLING!

Critics have called me an alarmist, a fearmonger, and a doomsayer for expressing my concerns about environmental issues—even though my concerns come from interviewing scientists all over the world and reading peer-reviewed research. So I couldn't help but shake my head at the kind of media coverage an industry report on the costs of the Kyoto Protocol received a few years back.

Some of it bordered on hysterical. One national Canadian newspaper, the *National Post,* ran a story with a banner headline across the top of the front page claiming that the cost of the Kyoto agreement to reduce greenhouse gas emissions was "pegged" at 450,000 lost jobs. How do they figure?

They figure by playing with numbers to fit a predetermined agenda. The "study" that got so much coverage (front pages and big articles in newspapers across the country) was a document from the Canadian Manufacturers and Exporters, an industry association. The report listed no formal author and was not peer reviewed. But you certainly wouldn't know that from the coverage it received.

The article in the *National Post* was particularly comical in its alarmist tone, reminiscent of old television commercials funded by the Global Climate Coalition (an American industry lobby group) where actors portraying "average" Americans complained that Kyoto would somehow cause old people to fall over in the dark and force everyone to drive tiny cars. The *Post* article left readers with the impression that the Kyoto agreement was some evil, foreign scheme plotted by unwashed anarchists under the dim glow of candlelight determined to push Canada back into the Stone Age.

The Kyoto Protocol actually came about only after thousands of scientists, researchers, and economists reviewed thousands of studies to come up with a reasonable course of action to start slowing global warming. Then politicians got involved and haggled over ways to make the transition easier for their countries. Canada pushed for many concessions to big industry and won most of them. Reaching agreement on the Kyoto Protocol was a long, comprehensive, and consultative process. It won't stop global warming, but it was an important first step.

Under the agreement, Canada was supposed to reduce emissions 6 percent below 1990 levels by 2012. The Canadian Manufacturers and Exporters report said unequivocally that meeting this target would require "radical" changes in our lifestyles: "We would all have to: Drive less, drive smaller cars . . . Reinsulate our homes . . . Pay up to 100 percent more for electricity . . . Pay more taxes." Oh, and a half-million of us would lose our jobs.

If reducing greenhouse gas emissions by 6 percent does all these things, then people in Toronto must all be unemployed, living in tents, and eating grubs by now. That city reduced municipal greenhouse gas emissions by 30 percent since 1990 and actually made money in the process. Being energy efficient actually saves money because it reduces waste. Why would that be a bad thing?

According to the industry report: "The issue of climate change is about global warming—it is not about air quality or smog." Oh, really? I suppose this strange statement allows the authors to free themselves from the fact that reducing greenhouse gases will also greatly reduce the air pollution that kills tens of thousands of North Americans prematurely every year and costs millions in health care dollars. The report also completely ignores the enormous costs that a changing climate will have on world economies—including Canada's. It ignores new business opportunities for efficient industries. And it ignores the savings to consumers who won't have to spend money on wasted energy.

Of course, we'll never know now how much we would have benefited had Canada and the U.S. actually lived up to their Kyoto commitments and worked to reduce greenhouse gas emissions. North America has fallen seriously behind the rest of the developed world in terms of energy efficiency. But it's never too late to catch up—so long as its citizens actually get moving and stop listening to those who want to hold them back. Ten years ago, big industry groups made headlines by saying that global warming didn't exist. After years of having their claims consistently proved false by scientists, they are now saying that global warming is a problem but slowing it will be far too expensive. If North Americans ignore them and get to work, ten years from now, those claims too will be exposed for the fearmongering that they really are.

.   .   .

MIXED MESSAGES OBSCURE
THE IMPORTANCE OF ACTION

Watching the news, it seems that there's a new environmental crisis occurring almost daily. Global warming, pollution, habitat destruction, species depletion and extinction . . . the bad news is plentiful and painful.

But then, on a fairly regular basis, a nice-looking man pops up and says that everything you are seeing is an exaggeration. The environment is fine. Everything's fine. Go buy a new SUV. It's okay. Such a nice-looking man.

To the average person, it must be quite confusing. What's going on here? Are the stories on environmental calamity really an exaggeration?

The quick answer is no. On a global basis, the environment we depend on for our lives is in trouble. Natural services that provide us with essentials like a stable climate, clean water, and fertile soils are being depleted. We are heading in the wrong direction if we want to leave the next generation with the quality of life and opportunities that my generation took for granted when we were young.

So why the mixed messages? Well, part of the problem lies in the way the media present news stories. First, they are presented as episodic, focusing on single events rather than issues and analysis. Second, media stories are driven by conflict, so reporters are always encouraged to find someone to contradict prevailing opinion, turning complex problems into a simplified, false "he says, she says" dichotomy. And third, there are well-financed interests at work who have a vested interest in maintaining the status quo, so they lobby hard to make sure their voices are heard.

Scientists have a duty to warn society of any potential environmental problems. But because of the incremental nature of science, not all predictions that are made will come to pass. Sometimes, action is taken to avert the problem, and sometimes the prediction turns out to be wrong. When that happens, it enables critics to say that there never really was a problem in the first place and environmental scientists are merely Chicken Littles who should be ignored.

To see if there is any truth to this argument, scientists at Princeton University and others conducted a study that looked at the costs and benefits of society's reaction to environmental alarms. Their report, published in *Science*, examines the costs and benefits of measures such as the Clean Air Act in the United States to see if society has truly benefited from these actions.

It has. The researchers found that regulation, for example,

has played the dominant role in improving air quality in the United States, earning Americans more than US$22 trillion in net benefits over twenty years. Furthermore, they found that in states or nations with equal wealth, those with higher memberships in "green" organizations and higher civil liberties have lower levels of air pollution. And they go on to point out that the costs of responding to environmental problems are often far less than originally anticipated.

The researchers conclude that society receives substantial benefits by responding to environmental alarms. In fact, they point out that "our environmental alarm is currently too conservative, not too liberal." In other words, far from being Chicken Littles, environmental scientists are perhaps being too cautious in communicating environmental problems. Certainly public policy-makers are slow to respond, as "Problems of detecting warning signals and overcoming vested interests inevitably lead to delay in regulation, often incurring damages that could have been prevented with higher sensitivity."

Critics of this report will no doubt say something to the effect that it's merely a case of alarmists supporting alarmists, but that argument amounts to nothing more than grasping at straws. The sooner we get serious about dealing with our environmental problems, the greater the benefits will be. True, it would be easy to listen to the nice lobbyists who tell us that everything is just fine, but waiting to take action will only make the good news less frequent and the bad news much worse.

.   .   .

CHANGING THE FUTURE MEANS CHANGING PRIORITIES

In 2004, news that global warming could push one-quarter of the world's plants and animals to the edge of extinction by

2050 made headlines around the world. But did the stories do more harm than good?

The forecast is truly grim. A major international research paper published in *Nature* reports that about one million species could be doomed to extinction. The culprit? Heat-trapping gases we are pumping into the atmosphere through vehicle exhaust, power plant and factory smokestacks, and home chimneys.

So, faced with such alarming news, what did people do? Did millions say "That's it, no more SUV for me!" and commit to public transit? Did thousands call, write, or email their elected leaders and demand action on global warming? Did corporate executives commit to making their industries cleaner?

No. Instead, this terrible news was largely met with a collective shrug. To be fair, it was more of a collective "Isn't that horrible!" or "How awful!" rather than a shrug, but the effect is the same. For the average person, nothing will change.

Why? Well, most people are simply overwhelmed by such news because the whole thing seems beyond their control. When that happens, people tune out. They have too many other things to deal with in their lives to figure out one more problem in the world—especially something as incomprehensibly big as climate change. Rather than being spurred to action, such news without context can drive many people to a defensive position.

Just as consistent news reports focusing on street crime create fear and drives people off the streets (thereby actually making the streets more dangerous), overwhelming environmental news also causes people to retreat into themselves and inadvertently make things worse. For example, rather than taking action to reduce global warming, some people may purchase a bigger, heavier vehicle "to keep the family safe."

Believe it or not, this actually creates a market for more environmentally unfriendly products. I can see automotive executives sitting in a boardroom discussing how to capitalize on concerns about global warming. "People are worried about global warming. What can we do about that?" asks an executive. "Bigger air conditioners to beat the heat!" says one. "Bigger everything to make people feel more secure," says another. This is what's called "meeting market demand."

Climate studies are vital to our understanding of the atmosphere and how we are changing it. But as studies pile up, we have to look beyond the impacts of what will happen if we don't change to how to make the changes necessary to prevent the worst-case scenarios from coming to pass. Right now, we are stuck in the mud and spinning our wheels.

All citizens can help by reducing the amount of energy we use. But to really slow global warming, North Americans need the kind of leadership and strategies that can only occur at the highest levels. The Canadian government adopted the Kyoto Protocol, the first small but important step to address climate change. Unfortunately, when the Conservative Party was elected, Prime Minister Harper quickly dismissed the very idea that Canada would even try to meet our Kyoto targets. In the U.S., President Bush also backed out of the deal and pushed for more oil drilling, rather than renewable sources, to meet energy needs.

With a new, clearly more thoughtful and visionary president now in Washington and environmental issues still in vogue, we can only hope that North America's federal politicians will stop being so shortsighted and follow the lead of many of our cities, towns, states, and provinces. Many of these jurisdictions have already made significant strides in the move towards sustainability. But they've all been crippled by a lack of vision and leadership from their federal leaders.

So, Mr. Obama and Mr. Harper—the ball's in your court. It's up to you to step up and catch up to the rest of the world, for the sake of the environment and our future. One million species and the next generation will thank you.

.   .   .

HIRED GUNS AIM TO CONFUSE

Al Gore once told me that to get politicians to listen, you have to engage the people first. The former vice president attempted to do just that with his critically acclaimed global warming documentary, *An Inconvenient Truth.* But he's up against some pretty powerful opponents.

His movie, by most standards, is pretty good. Rotten Tomatoes, a website that compiles movie reviews from newspapers, television, and the internet, shows that 92 percent of critics liked it. A story by the Associated Press on experts who critiqued the science behind the movie found that they, too, gave it a thumbs-up for accuracy. Personally, I thought it was brilliant.

But shortly after the Associated Press article came out, other articles started popping up that said Mr. Gore's science was shoddy. People claiming to be experts wrote opinion pieces in newspapers decrying the film, Mr. Gore, and the "theory" of global warming in general. Contrarians, it seemed, were coming out of the woodwork. What happened?

What happened was a well-funded campaign to discredit the film and carpet-bomb North Americans with confusing and contradictory information about the science of global warming. It appeared to have an effect, too. Polls done around that time indicated that while the public was very concerned about climate change, they were still confused about the science.

Those who read science journals probably find this public confusion, well, confusing. While there is plenty of discussion in scientific circles about what precisely a changing climate will mean to people in various parts of the world, there is no debate about the cause of global warming (human activities—mostly burning oil, coal, and gas) or about the fact that it is already having an effect and that those effects will become more and more pronounced in coming years.

Yet there they are in the editorial and opinion pages: supposed experts writing about the grand global warming conspiracy perpetuated by Europeans. Or socialists. Or European socialists.

Those in the know can laugh off such nonsense. But the problem is, most people aren't in the know. Average citizens are busy people, and they are not experts in climate science, so naturally they tend to defer to people who appear to know what they're talking about.

Unfortunately, masquerading as an expert in the media is pretty easy. All you need are a few letters after your name and a controversial story to tell. That makes news. And there's no shortage of public relations people willing to spin a good tale—usually for a tidy profit. Companies pay big bucks to have these spin doctors work their magic and make sure the industry line gets heard.

But even some of public relations' best-known spin doctors are disgusted by what's going on over global warming. Jim Hoggan is one. He's a personal friend who happens to be president of one of western Canada's largest public relations firms, James Hoggan & Associates. And he's so appalled at what he says is deliberate manipulation of public opinion about this issue that he started a website called DeSmogBlog.com to debunk the global warming skeptics.

Jim writes in his blog: "There is a line between public relations and propaganda—or there should be. And there is a difference between using your skills, in good faith, to help rescue a battered reputation and using them to twist the truth—to sow confusion and doubt on an issue that is critical to human survival. And it is infuriating—as a public relations professional—to watch my colleagues use their skills, their training, and their considerable intellect to poison the international debate on climate change."

Well said, Jim. His blog makes fascinating reading. It names names and follows the money trail—often leading back to big U.S. conservative organizations and fossil fuel giants. Jim's making it his mission to expose the liars and the frauds, and he's doing a pretty good job.

Al Gore was right: the people do have to be engaged before politicians will listen. But engaging the people sometimes requires clearing the air first.

.   .   .

NEWS GAME NOT ALWAYS COMPATIBLE WITH SCIENCE

"All seafood could disappear by 2050, new report," was the headline. But the psychological effect may as well have been: "Abandon hope, all ye who enter here."

Versions of the former headline abounded after a groundbreaking research article on marine biodiversity was published in *Science*. "Kiss your fish and chips goodbye" was another popular heading, as were takes on "No more fish in the sea."

On the one hand, this kind of alarming headline could be potentially beneficial because it highlights the urgency of a dire situation in our oceans. Without that sense of urgency, no one will act to prevent a disaster from occurring, and we really could lose most of our sea life. On the other hand, personally, such headlines make me want to bury my head in the sand or stick my fingers in my ears and sing choruses of "La la la, I can't hear you."

When news is so depressing and on such a huge scale, it can make individuals feel powerless. And when people feel powerless, they tune out. That's not how change happens.

Interestingly, the actual title of the research article published in *Science* was "Impacts of biodiversity loss on ocean ecosystem services." The point about the potential for catastrophic declines in sea life abundance was a relatively minor one in the study, used to highlight the urgency of the need to change the way we manage our oceans. The main thrust of the article was much more interesting.

That thrust was the importance of biodiversity in maintaining healthy marine ecosystems. The international study, headed by researcher Boris Worm out of Dalhousie University in Halifax, Nova Scotia, looked at a variety of marine ecosystems and how well they handled stress. It concluded that the more diverse an ecosystem, the better it is at dealing with stresses such as overfishing.

Biodiversity has long been seen as an important factor in the stability of land-based ecosystems. Biologically diverse ecosystems on land tend to be more stable. Compared to ecosystems with fewer species, land-based ecosystems with a greater diversity of life are better able to adapt to change, helping to secure the continued functioning of the entire system. This was the first comprehensive study to find the same is true for water-based ecosystems, discovering a consistent pattern across thirty-two small-scale experiments and through reviews of twelve coastal ecosystems.

The conclusion has major ramifications for the way we manage our fisheries, which still tends to be based on individual fish stocks rather than the ecosystems in which they are embedded. According to the new study, we're going about it all wrong. If you want to protect individual fish stocks, you really need to protect entire ecosystems.

Unfortunately, that story isn't very newsy. Disappearing seafood is. Without the news hook of the dire predictions for the future of seafood, the article may not have made the front page or any page at all in the popular press. So, either by chance or by design, the report's authors rolled out their study baited with the sweet smell of disaster.

And reporters took to it like sharks to a chum line, resulting in headlines around the world. Most newspapers and television stations stuck to the "total collapse" angle, often ignoring the biodiversity story altogether. More thoughtful journals, however, did focus on the actual thrust of the study—fisheries management and biodiversity. In its news pages, *Science* used the headline "Global loss of biodiversity harming ocean bounty," for example, while *The Economist* ran with "New research points to a better way of protecting fish stocks."

Whether the popular press stories were motivational or paralyzing remains to be seen. But the fact remains that right now, the spectacular and the spectacularly awful make headlines. In the news game, the rest is just details. That puts the way the mainstream press reports news at odds with the way people become motivated and makes social change even more difficult than it already is.

.   .   .

GOT A GOOD STORY? TELL SOMEBODY.

As a broadcast journalist, I'm well aware of the challenges today's reporters and journalists face in covering stories—from tight deadlines and a lack of resources to corporate ownership and the pervasion of tabloid-style reporting in mainstream media. But as guest editor for a Saturday edition of the *Vancouver Sun* newspaper, I found out that I still have a lot to learn.

I've never been a news reporter. In fact, more often than not, I'm the focus of a news story, not its writer. Still, I thought I had a pretty good idea of how the news game worked. I know that news is what's happening right now and that reporters have to crank out copy fast. And I know that daily news is an ephemeral beast. I myself have been guilty of picking up a newspaper, starting to read it, then throwing it down in disgust upon realizing that it was a day old. Yesterday's news just isn't news anymore.

So it was amazing to find out just how much goes into producing a daily newspaper. I was at the *Vancouver Sun* for a twelve-hour shift. Despite the fact that I had assigned some stories weeks before, there were still dozens of decisions to be made on the fly—everything from writing headlines to story placement, getting reporters to follow up on leads, use of language, fact-checking, and, of course, meetings, meetings, meetings.

And that was just the editorial part of the day. At 7:00 PM, when I thought we had put the paper to bed, we were off to the production facility where the paper was printed—another whole set of decisions and new challenges. The entire process left me exhausted and humbled.

Overall, I'm pretty pleased with the result. We managed to include some stories that I thought would never run—an article on the true cost of gasoline in the business section, for example. A reporter looked into what a liter of gasoline costs society if "full-cost accounting" is factored into the equation. This kind of analysis considers factors that are normally considered "externalities" in economics—things like air and water pollution and climate change. When these things are considered, gasoline actually costs upwards of $4.00 per liter ($15.16 a gallon)—far more than the $1.25 per liter or $3.80 per gallon we're currently paying at the pump.

My goal was to weave a common thread of sustainability throughout the stories. I hoped to get people thinking about the environmental footprint of everything we do and stimulate discussion about how we can do things better. It was actually pretty easy to find stories that touched on these issues for every section of the newspaper, from sports to arts. The reality is our economy and our way of life depend on the natural services that we generally take for granted. We can't afford to do that any longer.

I'm sure that some people were unhappy with "my" newspaper because I didn't make it only available online to save paper (great idea, but not an option for the publisher) or because car advertisements were still allowed in the paper, or because the stories weren't deep enough or didn't cover all the environmental challenges we face. In the end, it was just one day. I hope that the edition got a few people thinking in different ways. And I hope it gave the reporters and editors some new ways to think about things too.

So, here's my suggestion to everyone reading who, like me, gets frustrated with the media and the coverage of certain stories, or the lack of them: tell somebody. If you don't think your local newspaper, radio, or television station is covering something adequately, give them a call. Reporters are reporters because they are inquisitive people. They like telling stories. If you have a story idea, don't be afraid to write or call and suggest it. Environmental problems affect all of us. And it's up to all of us to solve them.

10

MR. SMITH GOES
TO **WASHINGTON**

Public policy for a sustainable planet

ASK MOST PEOPLE WHAT the number-one environmental problem in North America is and you'd likely get a variety of responses, from global warming to smog and from water pollution to species extinction. All of which are perfectly legitimate answers. But if you ask me, there's one that trumps them all: political apathy.

Why? Because sound public policy is key to making the U.S. and Canada greener and more sustainable. Enacting environmentally progressive government policies requires political will, which in turn requires public pressure. And when the public is uninterested in politicians or politics, there's no pressure—giving special-interest groups with deep pockets free rein to plan our future.

This hasn't necessarily been by accident. Analysts, pundits, and even some politicians themselves have warned the public so much about the dangers of government interference in our lives that it's become routine to oppose anything that looks like a regulation or a target. Big government is bad. Big government is scary. Big government is an expensive bureaucratic nightmare.

And it certainly can be. But in the press, "big" government has unfortunately become synonymous with "government that stands up to industry" or "government that does anything at all to help guide the future of the country." And that's just wrong.

Add to this attitude the fact that North Americans have suffered through a series of political scandals over the past few years, and it's easy to see why so many people have tuned out of politics altogether. In Canada, the recent federal election saw record-low voter turnout. To be fair, this was the third federal election for Canadians in just four years, so voter fatigue was certainly an issue. But in both Canada and the United States, the past decade has seen declining participation in the democratic process. Many North Americans would be hard-pressed to name their provincial premiers or state governors, their federal representatives, or their mayors. People have tuned out. They just want politicians to do their jobs and go away.

Until now. Finally, after more than a decade of apathy, there are signs of change. American voters turned out in droves to vote in November 2008. Some lined up for hours to cast their votes. The election of Barack Obama, the first black president, is another sign that Americans are embracing change and reengaging in the political process.

That's critical. For politicians to do their jobs, we have to do our job. Democracy is a privilege that too few people in the world enjoy. For our elected leaders to be able to change the status quo requires that they be given a mandate to do so by the people. If we elect governments that put a low priority on science or the environment, we can hardly expect them to enact policies that will put science or the environment first. But even if such a government is elected, voters still have power to change things. Public pressure can and does move mountains, but it needs to be vocal and it needs to be sustained because it's all too easy for politicians to ignore voices of change unless there is a good political reason to listen to them.

As we examine in this series of essays, there is no shortage of examples of the way governments ignore the advice of scientists, many of whom have actually been appointed as advisors by the government itself. Sometimes, the government has good reasons to ignore this advice, but most of the time it seems to stem from ideology or ignorance or both. We need look no further than our own backyard to see how a federal government can slow, stall, and even reverse decades of environmental progress to suit its own ideology. The Bush administration was as regressive in that regard as one can imagine. But American leaders do not hold a patent on political interference in science or on ignoring sound scientific advice. If Canadians are not vigilant as citizens, Canada could easily head down that same path.

Which brings us back to the apathy problem. No one wants a government that sticks its nose in our business all the time. But it is the responsibility of governments to set standards, goals, and targets for the betterment of all. That's the role of our leaders—to lead—and to ensure that the best interests of the country are served, rather than just those of an elite few. It's up to us, *all of us,* as citizens, to learn about the issues, figure out where we stand, and then tell our leaders how we expect them to behave.

When we allow ourselves to become completely disenfranchised, when we dismiss all politicians as "just a bunch of crooks" or say that "it doesn't matter who I vote for, they're all the same anyway," we're failing our country and we're failing democracy. Yes, well-funded industry groups and others will always have louder voices than we as individuals, but that's no reason to give up entirely. If you want to have a greener country, demand it from your president, your prime minister, your senators and MPs, your provincial premier and state governor, your mayor and councillors. Write them, call them, vote for them. Get involved, and start planning our future.

Because if you don't, the people with the deepest pockets will plan it for you.

.   .   .

WANTED: LEADERSHIP FOR
THE TWENTY-FIRST CENTURY

I just turned seventy-two. That's old—at least in my books. Sometimes I can't believe that I've made it this far. Other times I can't believe how much there is left I want to do. At my age, I think it's pretty common for people to start thinking about these things and what we want to leave behind—our legacies.

Politicians have a much shorter life span—politically speaking, that is. They can be around for four years or less. Rarely more than eight. That's why I'm often surprised by how little they seem to want to accomplish in that time. Certainly, I understand the lure of the status quo. Change is hard. Often vested interests will fight you every step of the way. Political advisors will say "No, no, no—stay the course! Don't make waves! Get reelected!"

But what's the point of being reelected if you aren't going to *do* anything? Yes, yes, maybe I'm being naïve. Maybe politicians are just there to support their vested interests, take home a fat paycheck and pension, and revel in the power of their office. But surely there's got to be more to it than that?

The life of a politician is not one I envy. It's hard, sometimes brutal. You are constantly under scrutiny. You are always on the job. It takes up your entire life.

That's why I honestly believe that most politicians at least start out wanting to work for the common good. Many become overwhelmed by the muck, but great leaders act. They make bold decisions and move on them. They don't tinker when big changes are needed, and they don't change things just for the sake of change. One of my pet peeves is the way some administrations will move into office and, rather than take an honest assessment of what's working and what isn't, instead set out to dismantle everything the previous administration has done just to make a point.

Of course, it's hard for leaders to act without public support. But right now, the environment is the top public concern. The public will support strong environmental leadership, so now's the time for our political leaders to act.

And politicians are indeed starting to take note. Seeing the success of initiatives in Europe, some politicians in North America are making bold decisions and plans to clean up our environment. Republican governor Arnold Schwarzenegger in California may have been the brunt of jokes when he was first elected, but no one's laughing now, as he's carefully crafted one of the world's most progressive legislative plans to reduce pollution and global warming.

Recently, B.C. premier Gordon Campbell went down to California to talk to Schwarzenegger about his plans. That's a very encouraging sign. Premier Campbell's Speech from the Throne was very bold and painted a new vision of British Columbia as leading North America in terms of sustainability. Given how proud British Columbians are of their natural heritage, progressive environmental leadership seems like a natural fit. It will also help diversify and strengthen B.C.'s economy in the long term and be a model for other provinces, states, and territories.

This is exactly what all our leaders should be doing—learning from each other. Many jurisdictions are coming out with exciting new programs to move towards sustainability. Ontario recently announced a "standard offer contract" system for renewable energy that's the first of its kind in North America. I hope Premier Campbell, and all our leaders, take a good look at the best examples of environmental leadership from all over the world and incorporate them into their own plans. It doesn't really matter where the ideas come from. Nature doesn't care about borders.

In the end, all that we have are our legacies. I've been on this planet now for seventy-two years. I don't know how many years I have left, but I promise you I plan to make the most of them. I hope our political leaders look at their terms in office the same way.

.   .   .

DON'T MISTAKE U.S. CRITICISM FOR CONTEMPT

What have you done? That was the question I wanted to ask my American friends after the 2004 election result. But I already knew the answer. They did everything they could. They just lost.

As a result, we were stuck with four more years of George W. Bush's regressive social, environmental, and foreign policies. At the time, the idea of another four years of W. seriously depressed me, and I wrote a column saying as much, noting that his reelection didn't bode well for science, the environment, or human rights in America—or elsewhere, for that matter.

Did this make me anti-American? According to many pundits and politicians weighing in on both sides of the border after the election, it did. Apparently, disagreeing with the U.S. popular vote made me either "anti-American" or "intolerant" or some sort of "high-minded liberal elitist." Even some of our elected parliamentarians in Canada insisted that any critical analysis of America or American policies simply amounted to "anti-Americanism."

The irony, of course, is that this was exactly the kind of "you're either with us or against us" mentality that drove many of the criticisms of the Bush administration in the first place. In his first term, President Bush forged a path of American unilateralism in the world community. He pushed a me-first agenda and was willing to trample human rights, science, and the environment to do it. Just ask the five thousand scientists, including forty-eight Nobel laureates, who signed on to a statement accusing the Bush administration of "manipulation of the process through which science enters into its decisions."

Yet after the election, those who dared criticize the choice of the slim majority of American voters who picked Bush were accused of being anti-American. Well, if being anti-American means being against the war in Iraq, supportive of women's rights, supportive of progressive environmental policies, against the missile defense system, supportive of stem cell research, and supportive of same-sex marriage, then sign me up. But I don't believe it does.

Simply disagreeing with that slim majority of voters does not make a person anti-American. In my youth I received a scholarship from an American university worth more than my father made in a year, and it allowed me to attend one of the finest colleges in the world. Later I earned a PhD there, and I am forever grateful to Americans for that. When I returned to Canada, I could not compete with my peers elsewhere in the world because of the poor funding available in Canada at the time. I stayed in my home country because I received a large U.S. grant. I will never forget the generosity of the U.S., and I owe a huge debt of gratitude.

But it is precisely because I love America that I was so profoundly disturbed by what was happening there. Unquestioning acceptance of the status quo isn't exactly an American ideal. In fact, it strikes me as decidedly un-American.

So yes, when 52 percent of Americans voted for Bush, I said that I thought they made a mistake. And when eleven states voted overwhelmingly to ban gay marriage, I spoke up. Disagreeing with a ban on same-sex marriage is not a matter of being out of touch with "American values." It's a matter of human rights. When one group in society is singled out and repressed and not given the same opportunities as others, then their rights are being violated. That is simply wrong. It doesn't matter if the majority of people voted for it. You can't vote away human rights.

Pundits who insisted that critics of President Bush were anti-American were really saying that if 52 percent of Americans believe anything, then that's what America stands for, and everyone else has to respect that. This is a morally relativistic viewpoint that doesn't even withstand the most basic scrutiny, and Bush administration critics should never have been bullied into believing it does.

Those of us who felt that 52 percent of American voters made a mistake on November 2, 2004, didn't hate Americans. On the contrary, we cared enough about the people and the ideals the country is supposed to represent to be very, very concerned.

LEARNING SCIENCE FROM PRESIDENT BUSH

With a Conservative prime minister in Canadian office, there's been plenty of talk about how much Stephen Harper will try to emulate American-style policies. We can only hope he doesn't follow the Bush administration's confused and confusing take on science.

Most people are probably aware of the official American government position on climate change, which has ranged from "it isn't happening" to "it may be happening but it has nothing to do with people" to "okay, maybe it is happening, but there's nothing we can really do about it." American diplomats even walked out of talks during the December 2005 climate negotiations in Montreal because they simply were not prepared to discuss any plans that would call for future reductions of heat-trapping emissions.

Meanwhile Tony Blair, the former prime minister of the country Mr. Bush called "America's closest ally," Britain, wrote in the foreword to the report *Avoiding Dangerous Climate Change* that "the risks of climate change may well be greater than we thought." Mr. Blair called the current growth in greenhouse emissions "unsustainable" and noted: "Action now can help avert the worst effects of climate change. With foresight such action can be taken without disturbing our way of life."

But back in the United States, the top climate scientist at NASA accused the Bush administration of trying to silence him. James Hansen says that after he gave a lecture promoting the necessity of reducing heat-trapping greenhouse emissions, the public affairs headquarters at NASA in Washington, D.C., ordered a review of all his upcoming lectures, papers, internet postings, and interview requests. Dr. Hansen told the *New York Times* that nothing in his thirty years with NASA compared to the scrutiny he faced in his daily activities under Bush.

And he's not the only scientist to complain about the Bush administration's unofficial policies of censorship. The Union of Concerned Scientists gathered a list of more than eight thousand researchers asking the White House to stop politicizing their disciplines. A preface to the list claims that: "Across a broad range of issues—from childhood lead poisoning and mercury emissions to climate change, reproductive health, and nuclear weapons—political appointees have distorted and censored scientific findings that contradict established policies."

Science should not be a partisan issue. In fact, perhaps the best criticism of the American government's politicization of science came from former Republican representative Sherwood Boehlert, who was also chair of the House Science Committee. In a letter to the NASA administration, he condemned the censorhip of Dr. Hansen, writing, "Political figures ought to be reviewing their public statements to make sure they are consistent with the best available science. Scientists should not be reviewing their statements to make sure they are consistent with the current political orthodoxy."

I could not agree more. I obtained my PhD from the University of Chicago in 1961 and shudder to think of what would have happened to my emerging discipline of genetics had the government of the time deemed that it did not conform to established policy. It's ironic that the government of a country that prides itself on innovation and the quality of its research would censor the same people responsible for America's continued technological and scientific success.

Fortunately, early signals are that Mr. Obama could have very different views on scientific research and environmental issues. He's spoken out in favor of stem cell research, for example, as well as the need for serious cuts to America's greenhouse gas emissions. Let's hope he nudges Prime Minister Harper in the right direction.

In regards to Mr. Bush's scientific and environmental legacy, if North America's two newest federal leaders can learn anything from his record, it's what not to do.

.   .   .

SQUIRREL SEX MAKES GOOD SCIENCE, DUMB POLITICS

A publicity stunt by an Ontario politician to tar his opponent for spending money on "squirrel sex research" may have made good media gossip, but it shows a shockingly poor grasp of science.

Ontario's conveniently named Progressive Conservative leader, John Tory, made front-page news a couple of years ago with his demand of Premier Dalton McGuinty to stop wasting taxpayers' money on flying-squirrel sex research. Calling it a "boondoggle" and "inexcusable" in a news release, he demanded that the premier rein in his "reckless" spending.

Well, I don't know much about the premier's fiscal management, but a quick look at the scope of this research finds that it's money well spent.

Contrary to Mr. Tory's claim, the research is not about sex habits, but rather "reproductive fitness"—that is, the species' ability to successfully reproduce. The study, conducted by Laurentian University professor Albrecht Schulte-Hostedde, was actually funded through an award that was originally set up by Conservative former premier Mike Harris for research excellence. The proposal went through several screening processes by independent experts and was also funded by the Natural Sciences and Engineering Research Council of Canada.

So did Dr. Schulte-Hostedde pull the wool over everyone's eyes? Were all these experts fooled by research that is surely very silly?

Hardly. Consider a lead paper published in the top-tier journal *Nature* in 2006: "Proteome survey reveals modularity of the yeast cell machinery." Not as funny as squirrel sex, but equally obscure. Perhaps it would have failed Mr. Tory's silly screen as well. But since when did politicians get to decide what makes good science?

Fortunately they don't. Or at least they shouldn't. In the case of the flying squirrels, the research is actually critical to helping understand how species are being affected by climate change. There are very few such long-term studies, and flying squirrels are a perfect candidate. They are an "indicator" species that tell us something about the health of the overall ecosystem. If climate change is harming the squirrels' ability to reproduce, then it's likely that other species are having difficulties as well. And that could have implications throughout the food chain.

Not all science has an immediate practical application. In fact, most of it does not. One of the biggest problems facing the future of science is the reduction in public interest and funding for basic research. Few researchers and even fewer funders are interested in basics that have little profit motive, like taxonomy, when the big money is in things like biotechnology and pharmaceuticals. Yet everything we know is grounded in basic research. If we don't cover the basics, we hobble our ability to understand our world.

Science does not progress in an easy, linear fashion. It's not like you have an idea, set up an experiment, prove your theory, and then cure cancer. In science, you learn as much from your failures as you do from your successes. Every paper, every theory, and every experiment builds on those that came before. As Sir Isaac Newton and other scientists have said—we're all standing on the shoulders of giants.

Political interference in science is a big problem. Despite global warming, the scientific climate in the United States has been pretty chilly for the past eight years. Scientists there accused the Bush administration of censorship, of fiddling with findings, and of hindering research. Is this the sort of thing we'd like to import into other countries? Should people who have never peered into a microscope decide what "good" science is? I think not. Research independence is critical to the advancement of science. If it were left to the flavor-of-the-month whims of politicians, we'd still be in the Dark Ages.

.   .   .

LABEL PLANETS, NOT PEOPLE

During the same week of February in 2006, Canadians elected a new prime minister and scientists announced the discovery of a possible "Earth-like" planet in a distant solar system. Observers joked that both were cold and distant bodies, but I think there are reasons to be more hopeful.

Scientists labeled the planet OGLE-2005-BLG-390Lb. At 5.5 times the size of our Earth, it is the smallest extrasolar body ever discovered, and it takes ten years to orbit its sun. It's also very, very cold—so cold it is unlikely to harbor any sort of life. Not exactly the kind of place we could call home.

Knowing that researchers are probing our universe, trying to understand where we came from, where we are going, and how it all works is somehow reassuring. It's heartening to think that one day in the not-too-distant future, we may discover that we are not alone in the cosmos, that Earth is not unique.

But we also have to remember that even if we do find other planets capable of supporting life, they are going to be a long,

long way from here. For us to be able to visit or "emigrate" to another Earth-like body would take a quantum leap in our technology and in our understanding of the universe. This type of trip simply isn't feasible based on current science. For now, such expeditions still belong in the realm of science fiction.

In the meantime, while we scan the heavens for other Earths, we have to remember to bring those curious eyes back down here, to the only Earth we know actually exists. As far as we can tell, it is unique—the only spark of life in a very cold and dark universe. And right now scientists are telling us that we are doing a poor job of taking care of it.

Which brings me to Canada's prime minister. When Stephen Harper was elected, I heard people comment that it was a disaster for the environment. After all, Mr. Harper has been labeled a "neoconservative," a "libertarian," and "anti-environmental." It hasn't helped that he has spoken out against even the most broadly supported of environmental initiatives, such as the Kyoto Protocol.

Now, I never minded being called an "environmentalist." In fact, I wore it as a badge of honor because it represented some of my core values. But the older I get, the more I realize that most people don't like being labeled. They don't like being typecast or pigeonholed. Labels are exclusionary by their very nature, and they push people to their ideological corners like boxers ready to come out for a fight.

The fact is, for every declared "environmentalist" there are a thousand people who care profoundly about the environment. And there are another ten thousand who recognize almost intuitively that human health and well-being are intimately connected to the world in which we live. As a leader and as a father, Stephen Harper, I am sure, recognizes this connection too. Polls show that the vast majority of Canadians

certainly do. I just hope Mr. Harper can see past his own ide-
ologies and make the necessary changes to put Canada back
on track.

The greatest environmental victories of the future will not
be made by environmentalists, but by millions of concerned
people taking small steps towards a common goal and ensuring that governments help us take those steps. These people
have no labels. They may be Liberals, Conservatives, New
Democrats, Bloc Québécois, Greens—it doesn't matter. They
simply see that taking care of our environment means taking
care of us. It makes sense for our health. It makes sense for
our economy, and it makes sense for our well-being.

The question we have to ask ourselves is, what do we want
our country to look like? Not just today and tomorrow, but
twenty-five, fifty, and one hundred years from now? What do
we want our grandchildren and great-grandchildren to thank
us for doing? When we look ahead to our long-term goals, we
can find common ground. And that's where we have to begin.

*   *   *

GUIDED BY UNCONSCIOUS VOICES

Attention all politicians: take my advice—sit on it. At least,
that's what the latest research is saying helps lead to the best
decisions.

While listening to their instincts or their "gut reaction" has
long been cited by people as a reason for making choices, scientists have often dismissed this seemingly irrational process
as merely "folk wisdom." Now science is catching up to that
age-old wisdom.

According to a report published in *Science*, complex decisions are best handled by the unconscious mind. Researchers

at the University of Amsterdam conducted a series of experiments on purchasing decisions and found that, while conscious deliberation is good for making simple decisions, for more complicated choices it's often better to sleep on it and then simply go with your gut.

Conscious thought does not always lead to the best decisions because it has low capacity. We can only consciously think of a small number of things at any given time, which can lead us to focus on minor details or only a small subset of relevant information. And because we can only focus on a small number of details at once, we are not very consistent with multiple evaluations of the same choices, since we may choose to focus on different attributes each time. This leads to what has been called "option paralysis" because we keep coming to different conclusions.

Our unconscious mind, however, is capable of integrating large amounts of information, although with less precision. For the Amsterdam university experiment, researchers postulated that because conscious thought is so precise, it would lead to good decisions over simple matters, where only one or two attributes were involved. However, because unconscious thought has such high capacity, that would lead to better decisions over more complex matters. They called this hypothesis "deliberation-without-attention" and they tested it using four studies on consumer choices, including some in a laboratory setting and some using real shoppers.

One study involved participants reading positive and negative information about four hypothetical cars. Some were then asked to think about the cars for four minutes and choose one. Others were distracted with another task for four minutes before making a choice. While those who had time to think about their choice made good decisions when the information was simple (only four attributes listed per vehicle), they more often made poorer choices when the information became more complex (twelve attributes listed per vehicle).

In another study, shoppers were quizzed about their purchases upon leaving an IKEA store and a department store. They were asked specific questions about the cost of their items, how much they had known about them before coming to the store, and how much time they thought about their purchases before they bought them. Follow-up phone calls revealed that shoppers who spent more time deliberating about simple purchases (such as kitchen accessories) and less time deliberating about more complex purchases (such as furniture) were ultimately more satisfied with their choices.

In his book *Blink,* former *New York Times* science reporter Malcolm Gladwell wrote about similar properties of the unconscious mind. Gladwell talks about "thin slicing," which refers to the mind's ability to find patterns in certain situations based on very narrow slices of experience. According to Gladwell, what some people call intuition or that "gut" feeling is really grounded in logic and shaped by our knowledge and experience with the world—we just aren't necessarily able to easily articulate it.

The Amsterdam researchers point out that, though they focused on consumer products, "there is no a priori reason to assume that the deliberation-without-attention effect does not generalize to other choices—political, managerial, or otherwise."

So to all our politicians: deliberate, debate, discuss. Then go home and sleep on it before making a decision. We'll all be better off if you do.

·   ·   ·

GROWING ENVIRONMENTAL CONCERNS
WORRY BIG INDUSTRY GROUPS

You might think, given all the media attention global warming has received and the success of Al Gore's *An Inconvenient Truth,* that everyone has finally accepted the problem and the need for change.

If only.

Standing in the way of reality are America's big "free-market" think tanks and the money and political clout behind them. These organizations, funded largely by big industries, come up with all sorts of unusual claims and statements to justify the status quo and make sure that big, profitable, polluting industries continue to be big, profitable, polluting industries. One of the most outrageous comes from the Competitive Enterprise Institute—two 30-second television ads that are so over the top, they almost parody themselves.

In one ad, over beautiful scenes of children playing in a park, a woman's soothing voice intones: "There's something in these pictures you can't see. It's essential to life . . . The fuels that produce carbon dioxide have freed us from a world of backbreaking labor. Now, some politicians want to label carbon dioxide a pollutant. Imagine if they succeed. What would our lives be like then?"

Another ad attacks the media and environmental groups for being "alarmist":

> You've seen those headlines about global warming. The glaciers are melting. We're doomed. That's what several studies supposedly found. But other scientific studies found exactly the opposite. Greenland's glaciers are growing, not melting. The Antarctic ice sheet is getting thicker, not thinner. Did you see any big headlines about that?

Why are they trying to scare us? Global warming alarmists claim the glaciers are melting because of the carbon
dioxide from the fuels we use.

Unbelievable. "Several studies." "Supposedly found." My
favorite line is read over a shot of some poor soul riding a bike
in a blizzard: "Let's force people to cut back, they say." Each of
these ads ends with the clever tagline: "Carbon dioxide. They
call it pollution. We call it life."

The second ad shows glimpses of scientific papers about
ice sheets, which speed past quickly enough that most people would just assume they must bolster the commercial's
argument. But if you track the papers down, you'll find the
opposite is actually true. The first paper points out: "There is
nonetheless considerable evidence of melting and thinning in
the coastal marginal areas in recent years." The latter notes
that: ". . . these observations are consistent with the latest
IPCC prediction for Antarctica's likely response to a warming
global climate." Not exactly smoking guns against the global
warming case.

Environmentalist groups have long been castigated for
being alarmist about global warming, so seeing such alarmist rhetoric from the opposite camp is quite ironic. But what's
driving industry groups to make these ads now? A review of
the film *An Inconvenient Truth* in *Nature* called the ads "arguably hilarious" and noted that they were plainly a response to
the publicity Al Gore received for his film.

That may be true, as Mr. Gore's film received wide
acclaim. But more generally, the ads are probably a response
to the growing concern about global warming and other environmental problems that are not being adequately addressed
in either the United States or Canada. President Bush barely

acknowledged that there was a problem at all, and Prime Minister Harper risked Canada's international reputation by refusing to live up to our Kyoto pledge.

People are starting to wake up to the tremendous challenges and dangers posed by unchecked greenhouse gas emissions. As they wake up, they may start to demand action from their political leaders. But don't expect the industry associations to accept that without a fight—even if that means using lies and deceit to confuse people about one of the most critical issues of our time.

.   .   .

WHAT WOULD YOU DO IF YOU WERE PRIME MINISTER?

I love Canada. Its peoples. Its geography. Even the weather. It's a good thing, too, as I experienced it all up close and personal on a cross-country adventure to talk to Canadians about the environment.

Throughout the month of February 2007, I made stops in more than forty communities, from St. John's, Newfoundland, to Victoria, B.C. It wasn't a book tour or a publicity tour for a television show. It was something I'd wanted to do for a long time—start a conversation with Canadians about our environment, our children, our grandchildren, and our future.

I believe there is a fundamental disconnect between Canada's elected leaders and its people. Polls tell us that environmental issues like global warming are the number-one concern of Canadians. Yet most of its politicians offer up little more than window dressings to address these issues. It's as if many are just hoping to lay low until this "environment thing" blows over so they can go back to ignoring it as usual.

That's not right. And I personally will do everything I can

to make sure that doesn't happen. I want to make sure Canadians' concerns are heard in Ottawa.

No matter what your political stripes, we all depend on a healthy environment. Brian Mulroney was recently voted Canada's greenest prime minister, and he's a Conservative. Whether he was really interested in the environment is debatable. But the fact is, he had no choice but to go green, because the public demanded it.

In the late '80s and early '90s, environmental issues were hot. Even George Bush Sr. was elected by saying he would be an "environmental president." Corporations and governments set up new environment departments and started "going green." Recycling was all the rage.

In the public eye, the problem looked like it was solved. People were recycling. Governments consistently talked about the importance of the environment. Corporations shined themselves to a deep green luster. People breathed a sigh of relief and went back to their everyday lives. Unfortunately that green luster was only skin deep. Beneath the surface, little had changed.

As a result, we essentially went on with business as usual. And it wasn't until global warming started being actually observed by people and reported on by the media consistently and accurately that the environment got back on the agenda.

That's where we are now. Only this time, if we want to actually move our countries towards a cleaner, healthier, and more sustainable society, we have to do more than just brand positioning and image makeovers. We need real change. We need to have strong targets and timelines for our biggest polluters to reduce their greenhouse gas emissions. We need to clean up our cars, our homes, and our businesses. We need to build sustainability into the bottom line.

If you want to make your country greener and more sustainable, get involved and tell your elected leaders that it's not good enough to just smile and nod for the cameras. Tell them that you expect more.

Although Prime Minister Harper has made it clear that the environment isn't a priority for his government, if enough Canadians demand change, he will have to listen. With a new president in Washington who appears to be more open to environmental, scientific, and social issues than was President Bush, Mr. Harper no longer has an ideological ally south of the border. Faced with internal and external pressures, Canada can only hope that Mr. Harper changes his tune, because our planet deserves better.

.   .   .

TAKE MY RESEARCH, PLEASE

Not long ago, news blogs and newspapers reported that some politicians had cribbed research conducted by my foundation and used the information to build their own environmental agendas. This news sent many a blogger all atwitter. While some of them focused on whether or not the information had been adequately referenced, others decried this action on the part of the politicians as proving that they had no ideas of their own, so they had to steal them from others.

Allow me to clear something up right now. To all politicians looking for ways to reduce our footprint on nature—or, to use politician-speak, create an "environmental platform": knock yourselves out. Feel free to steal, pilfer, borrow, rent, filch, or otherwise take any research my foundation does and put it to good use.

This may seem obvious to some, but the whole point of conducting and publishing this research is to get people to actually use it. As public education, it helps raise awareness of environmental problems. But more important, it provides solutions to those problems. And most of those solutions are best implemented by our political and business leaders, rather than by individuals.

So if you ask me if it bothers me that politicians are stealing the solutions brought forward by my foundation, the answer is no. To use a computer term, we consider this information "open source." It's a free buffet; please take all you like. The whole reason we do the research is to effect change. If those who have the power to make those solutions happen actually use that information, so much the better. This is how change happens.

As for the complaint that using my foundation's ideas shows that politicians have none of their own—nonsense. Since when do great leaders come up with all their ideas on their own? Societies built around the narrow viewpoints of one person are called dictatorships and tend to be decidedly backward and not terribly pleasant. And if the notion is that ideas should be coming only from within a particular party— again, nonsense. This kind of partisan mentality is a form of xenophobia, and it kills new ideas. Then again, perhaps that explains the state of Canadian politics.

I've also been asked if I worry that if one political party "steals" our ideas and runs with them, it might be off-putting to the other parties. That is a concern. But we can't control who uses our research, and we don't want to. The David Suzuki Foundation is non-partisan. We share our research with all political parties and encourage them all to adopt the solutions we bring forward.

Frankly, it's a tough slog all around. We can have a great idea and support from the vast majority of the public, but political leaders can turn it down flat because it might cost

votes in an important constituency or because of political lobbying from an industry group. Sometimes there doesn't appear
to be any reason why an idea is rejected other than fear of
change. That can be disheartening, but at least if the information is out there, the public can use it to make changes in
their own lives or to ask our leaders to take action.

My foundation is just one of dozens of organizations across
Canada and hundreds across North America offering solutions to our environmental and social problems. Rather than
ignoring these solutions because they don't come from within
a particular party, it is my hope that our political leaders open
their eyes, embrace change, and start taking advantage of all
this free advice. That isn't stealing; it's just good leadership.

.　.　.　.

GREEN IS IN. LET'S KEEP IT THERE.

Two years ago, if you had told me that today we would be riding a wave of environmental awareness not seen for nearly
twenty years, I probably wouldn't have believed you. Yes, 2006
had *An Inconvenient Truth,* but since then we've seen an
explosion of environmental concern.

As great as this has been, we mustn't forget that news is
a fickle beast and, by definition, "new" doesn't last very long.
That means we need to keep the interest moving forward, or
we could lose the momentum we've built up.

We've come a pretty long way in two years. In 2006, people started paying attention to the environment again. It was
like society woke up from a collective environmental slumber,
looked around with bleary eyes, blinked, and asked, "What's
going on?" And people started to look for answers.

In 2007, the media got on board. Environmental stories made front-page news all year long. Books about the

environment became best sellers. Magazines on topics from home design to celebrity gossip suddenly had environment pages or "green" tips. Eco-this and enviro-that became commonplace. My local newspaper, the *Vancouver Sun,* actually invited me to be guest editor and has seen a newfound interest in environmental stories. Until the economic collapse in 2008, the environment was the top-of-mind issue for most people. Even after the markets tanked, concern still remained high.

But for those of us old enough to remember back a couple decades, this might seem like déjà vu. In the late '80s and early '90s, the environment was also a top public concern. Governments poured money into environment ministries. Corporations developed environmental stewardship platforms. Municipalities across the country rolled out blue-box recycling programs.

And the people cheered. Problem solved! Now they could go back to worrying about more manageable individual priorities like paying bills, going to work, and providing for their families. Of course, we all know now that the problem wasn't solved by a long shot. But we lost a decade of potential progress because people slipped into complacency. And who can blame them? No one wants to deal with something as big and complicated as our global environment.

The thing is, we don't have a choice anymore. Leading scientists have been telling us for decades that we are on a very dangerous path. The good news is that it is not too late to change the route we are on. There are alternative ways to live that are in balance with Earth's life support systems. But getting on a new path requires real change.

So what does "real change" mean? For governments, giving money to an environment ministry only to have its mandate trounced by the ministry of natural resources, energy, or finance won't cut it. For corporations, token efforts like a "green" building design or energy-efficient lighting won't cut it if your bottom line is still profits at the expense of the environment. For individuals, using reusable bags instead of plastic or carrying a reusable coffee cup will do little if you're still driving to work every day.

If it sounds like a challenge, that's because it is. Real solutions are never easy, and there will be lots of arguments. We will also make mistakes. But the only real failure will be if we don't try at all. We've only got one Earth, so we can't very well just wait and see what happens if we continue with business as usual. That path may look easy now, but in the near future it will make things very, very hard.

So what does this all mean?

It means it's time to dig deeper.

We already have the public's attention, but now we need to get serious about solutions. We know how hard it is to be environmentally responsible. Many of our daily decisions are not good for the environment, because it is easier and often cheaper for them not to be. Still, challenge yourself, challenge your neighbors, challenge your friends and coworkers to make your city or country a global warming problem-solver, not a problem-maker.

Large-scale changes also require corporate and government leadership. But here, too, individual action can have great power. Politicians and business leaders know the public is concerned, but they are slow to respond unless really pushed. If you really want to make a big difference—push them. Push them hard. Change is underway. Real change is happening. Let our leaders know there's nothing that can stop it.

FINAL WORDS

TEN YEARS MAY NOT even register on a geological timeline of the history of our planet, but it is a long time in terms of how fast science and technology are advancing. Yet, looking back, we find that most of the issues we were dealing with a decade ago, we are still dealing with today. This tells us two things.

First, the difficulties we're encountering in solving our environmental problems aren't scientific or technological; they're social. As some of these essays point out, many scientists and researchers agree that we already know what must be done to achieve sustainability and even how to do it; we just aren't acting.

Second, relying on science or technology to save us from our environmental problems is a fool's game, because both science and technology are reflections of society. If we don't commit as a society to change, science and technology will get us nowhere and in fact will take us even faster down an unsustainable path, because that's where we're already heading. ·

But if we do commit to change—that's when things get really interesting. Imagine harnessing all the power of science and technology for the good of humanity. Imagine including environmental health as an indicator of economic well-being. Imagine the cost of polluting goods and services actually reflecting the damage they cause to human health and the environment. Imagine proactive environmental policies designed to prevent environmental damage from occurring in the first place, rather than simply trying to clean messes up later. All of this is possible, but only if we as individuals, as a society, and ultimately, as a species decide that this is what we want.

This means that it's no longer good enough to shrug our shoulders at the world's problems and say: "What can you do? That's the price of progress!" or "No use fighting it—it's human nature!" or merely daydream about technologies that could one day save us from ourselves. It's no longer good enough because these things are up to us. The choices we make determine which areas of science will be studied, which technologies will be adopted, and which public policies will be enacted. Choosing to let these things wash over us as though we are just impartial observers or helpless victims of humanity's vices is still a choice. Each of us is responsible, in some small way, for the future of our species.

Humanity has been called a scourge on the planet, a parasite, even a cancer. I think none of these things. We are a unique and wondrous creature. After all, no other species that we know of in the entire universe is even capable of having this kind of discussion. No other species is capable of sharing with us its own unique experiences and histories. We're it. We've done great things and are capable of much, much more.

But ultimately, you take away all of our scientific knowledge and all of our technologies, all of our art and all of our culture, and we are still just another species. We need air to breathe, water to drink, and food to eat just like every other organism on our little planet. We are but a small, shiny cog in the big wheel of life on Earth. And as shiny and as fascinating as we are, we need the humility to recognize that fact, because in the end, the Earth doesn't need us.

But we certainly need it.

ACKNOWLEDGMENTS

THE BIG PICTURE WOULD not have been possible without the continued and tireless efforts of the staff of the David Suzuki Foundation. From this remarkable group of individuals has emerged a voice for change in Canada—and across the world—that is as visionary as it is compelling. We tip our hats to all their work.

Special thanks go to Foundation staff members who had a hand over the years in helping with these essays, from suggesting story ideas to research and fact-checking. Dr. Faisal Moola, David Hocking, Ann Rowan, Dominic Ali, Autumn O'Brien, Teresa Laturnus, and many others all made sure Science Matters was ready on time every week.

To Jim Fulton, thank you for inspiring us with your energy and enthusiasm and for showing the true meaning of determination when the stakes couldn't be higher.

To Dr. Tara Cullis, thank you for your continued support and for providing a name for the column almost a decade ago.

To Cris Leykauf, thank you for wisely telling Dave that it wouldn't be the end of the world if he left journalism to become a communications flack if he believed in what he was doing. Thank you, too, for bringing Parker into the world and giving Dave his most powerful reason to protect it.

Thank you Robert and Joyce Taylor for inspiring Dave's love for the natural world and for leading by example.

Finally, to Nancy Flight and Rob Sanders at Greystone Books, our continued thanks for your efforts over the years to ensure that stories about science and the environment continue to receive the attention they so deserve.

INDEX

advertising, 2, 91–92; anti-regulation, 150; by auto industry, 148; to children, 205–7; and climate change, 262–64. *See also* media
agriculture, 177–79. *See also* farming, organic
air pollution, 27, 149–50, 196–98
animal studies, 10–12
anti-Americanism, 250–52
apathy, political, 245–48
Aral Sea, 52–57
automakers, 140–41, 148–50, 150–53, 160–62. *See also* cars
awareness: environmental, 221–22; scientific, 26

bats, 47–49
biodiesel, 158–60
biodiversity: hotspots, 77–78; importance of, 59, 83; marine, 241–42; understanding, 60–62, 213
biofuels, 158–60
biophilia, 59, 81, 199, 212–13
biotechnology. *See* GMOs
birds, 63–65, 178–79
birth control pills, and fish, 44–45
Blair, Tony, 253
Blink (book), 261
Boehlert, Sherwood, 254
boreal forest, 104

brown clouds, 134–35
Browne, John, 125
Bt, 180–81
"bubble fusion," 15–16
Bush, George W.: criticism of, 251; and environment, 247; and Kyoto Protocol, 95; reelection of, 250–52; and science, 226, 251, 253–55, 257
butterflies, 65–67, 180

Campbell, Gordon, 249–50
Canada: and automakers, 160; and birds, 65; and climate change, 37, 42; and GMOs, 185; and Kyoto Protocol, 96, 126; sustainability in, 42; wetland draining in, 90
captive breeding, 67–69
captivity, animals in, 10–12
carbon dioxide. *See* greenhouse gases
carbon taxes, 107–9
cars, 136–38, 139–40. *See also* automakers
CFCS, 51, 96
children: advertising to, 205–7; curiosity of, 26; environmental education, 214–16; and nature, 205–6; and science, 26; and technology, 207–9; and touch, 201–2

Christmas, 216–17
Chrysler. *See* DaimlerChrysler
climate change: and advertising,
    262–64; and Canada, 37,
    42, 126; economics, 92–93,
    100–102; effects, variety
    of, 117–19; extinctions, 83;
    feedback loops, 118–19; and
    food supply, 115–17; and forest
    protection, 121–24; and global
    dimming, 27; inaction on,
    124–26; and media, 219–20,
    224–26; perceptions of, 110–12,
    112–14, 129–31, 131–33,
    238–40; and plants, 113–14,
    117–18; public interest in,
    133–35; responses to, 235–38;
    scientific consensus on, 128;
    skeptics, 28, 111, 129–30; and
    United States, 37, 42, 126, 253;
    urgency of, 51, 57; and weather,
    119–21. *See also* Kyoto Protocol
conservation medicine, 47–49
conservation of biodiversity, 60–62
consumer products, 91–93, 203–5
consumers, and environmental
    change, 139–41
Costanza, Robert, 90, 95
crop yields, 169–70

DaimlerChrysler, 140–43, 143–45
David Suzuki Foundation
    research, use of, 266–68
decision-making processes, 259–61
DeSmogBlog.com, 239–40
diclofenac, 63–64
dioxins, 51, 170–71
diversity, 61. *See also*
    biodiversity
Dixella suzukii (fly), 81–83
DNA, 19–21, 22–23, 32
Dobson, Andrew, 49
Dobzhansky, Theodosius, 32
Drake, Frank, 18

Earth, 36, 55, 56, 57
Earth-like planets, 18–19, 257–58
ecological debt, 105–7
"ecological footprint," 105
economic growth, 93–95
Economist, The (magazine),
    104, 242
economy, 3, 94
ecosystem services, 13, 39;
    economic valuation of, 62,
    84–86, 89–91, 102–5
education, 31–33, 214–16
elk, 71
ENCODE project, 22, 23
environmental awareness, 268–70
"environmentalist," as label, 257–59
estrogen, and fish, 44–45
ethanol, 158–60
evolution, 31–33
externalities (economic), 93,
    98, 192, 243–44
extinctions: and agriculture, 179;
    birds, 63; butterflies, 66; and
    climate change, 83, 235–36;
    golden toad, 83; on islands, 78;
    mass, 13, 59, 66; plants, 66;
    preventing, 77–79; risk of,
    63–65. *See also* species: loss of
extraterrestrial life, 17–19

Fangmeier, Andreas, 116
farmed salmon, 170–71, 172–73
farming, organic, 166–68, 168–70,
    182–84. *See also* agriculture
fathead minnow, 44, 45
fish, 44–45, 173, 174.
    See also salmon
fisheries, 175–77, 240–42
flying squirrels, 255–56
food, 115–17, 163–65, 186–88,
    189–91, 191–93; organic, 223–
    24 (*see also* farming, organic)
Ford, Bill, 140
forest protection, 62, 104, 121–24

Frankenstein, Victor, 8–9, 10
Frankham, Richard, 68–69
fuel efficiency, 140, 147, 150–53
full-cost accounting, 192, 243–44
Fulton, Jim, 151–53

Gaia hypothesis, 45–46
gasoline, true cost of, 244
GDP, 87, 94, 99
Gladwell, Malcolm, 261
global dimming, 27–28
Global Footprint Network,
     105, 106
global warming. *See* climate
     change
GMOs (genetically
     modified organisms), 8,
     28, 179–82, 184–86
golden toad, 83
Gore, Al, 126, 238, 240, 263
Great Bear Rainforest, 75–77
greenhouse gases, 51, 57,
     102, 122, 262–63
grizzly bears, 37
Guatemalan howler monkey, 78

habitat loss, 90
Hansen, James, 224–26, 253–54
Harper, Stephen: and environment,
     258–59, 264–66; and Kyoto
     Protocol, 95–96, 97, 126, 237,
     258–59; and science, 253, 254
heat waves, 113–14, 134
Hoggan, Jim, 239–40
hormone disruption, 37, 43–45
hotspots, of biodiversity, 77, 82–83
human genome, 19–24
humans: adaptations by, 35–36;
     and cars, 136–38; connections
     with each other, 43, 58, 217–18;
     connections with nature, 43, 73,
     213; and food, 163–65, 189–91;
     history, 208–9; perception of
     biodiversity, 213; population,

61, 177–78; and ritual, 216–18;
     and technology, 194–96;
     uniqueness of, 272
Hurricane Katrina, 126–27, 128
hybrid cars, 139–40
hydrogen economy, 155–57

Inco, 96–97, 149
intelligent design, 31–33
interconnections: of humans,
     43; in nature, 6, 34, 36, 39,
     45–46, 47, 55
Irwin, Steve, 79–81
islands, extinction risk on, 78

journalists, 24–25, 31, 220,
     242–44. *See also* media

Kennedy, Donald, 33, 124
Kermode bears, 75
Kilimanjaro, Mount, 142–43
killer whales, 12, 37
krill, and ocean mixing, 46–47
Kyoto Protocol, 95–96, 97–98;
     and Barack Obama, 237–38;
     economics of, 97, 125, 231–33;
     and forest protection, 121–24;
     and George W. Bush, 95;
     and media, 225, 231–33;
     and Stephen Harper, 95–96,
     97, 126, 237, 258–59.
     See also climate change

Lander, Eric, 21–22
latent hotspots, 77–79
leadership, need for, 248–50, 270
LEX scale, 25
logging, 122
Lomborg, Bjørn, 229, 230
Lovelock, James, 45

MacGregor, Roy, 41
Maddox, Sir John, 5
Mazhitova, Zita, 54

meat production, 170–72
media: and "balance," 220–21;
    and Bjørn Lomborg, 230;
    and context, 2; and
    environment, 233–35, 268–69;
    and fisheries collapse, 177;
    and framing, 27; and Kyoto
    Protocol, 225, 231–33; and
    science, 20, 24–26, 29–31;
    and social change, 219–22.
    See also advertising, journalists
microorganisms, 13–14
Millenium Ecosystem
    Assessment, 104–5
misinformation campaigns, 29–31
Montreal Protocol, 50–51
Mooney, Chris, 30
Myers, Ransom, 175–76, 177

NASA, 19, 253, 254
natural selection, 68
nature: and children, 205–6;
    conservation of, 41–43;
    economic valuation of, 86–89
Nature (journal), 30, 230
New Scientist (magazine), 130–31
nitrogen, 66
northern cod, 176
nuclear fusion, 15–17

Obama, Barack, 237–38, 246, 254
obesity, 153–54
objectivity in science, 9
obsolescence, planned, 209–11
ocean turbulence, 45–47
option paralysis, 260
orcas, 12, 37
organic farming, 166–68,
    168–70, 182–84
outdoors, health benefits of, 212–13
ozone layer, 50–51, 132

parks, 72–74, 207–8, 213
PBDES, 172–73, 174–75

PCBS, 37, 51, 170–71, 173
plankton, 14
plants, 66, 113–14, 116, 117–18
poison ivy, 117–18
polar bears, 37, 70
political apathy, 245–48
politics: and science, 16,
    30, 253–55, 255–57.
    See also names of politicians
population, human, 61, 177–78
poverty, 98–100, 226–28
predators, 70–72, 175
primates, 39–40, 69

rabies, 48, 64
rainforests, 74–77
Reducing Emissions from
    Deforestation and Degradation
    (REDD), 123–24
reductionism, 6, 9–10
regulations, 95–98,
    197–98, 234–35
Republican War on Science,
    The (book), 30
rice, arsenic in, 38
ritual, importance of, 216–18
RNA, role of, 23, 24
Rodale Institute Farming
    Systems Trial, 183
rodents, and memory, 11
Roszak, Theodore, 8–9

salmon, 34, 37, 69; farmed,
    170–71, 172–73
Schrag, Daniel, 125
Schulte-Hostedde, Albrecht, 255
Schwarzenegger, Arnold, 249
science: ambiguity of, 26–29,
    128–29; awareness of,
    importance, 26; benefits and
    risks of, 5–7, 8–10; education,
    31–33; flexibility of, 7, 16; and
    hope, 16; incremental nature
    of, 7, 128–29, 234; information

sources about, 222–24; and
  LEX readability scale, 25; in
  media, 20; objectivity of, 9, 29;
  perceptions of, 24–26; and
  politics, 16, 30, 253–55, 255–57
Science (journal), 242; and
  "bubble fusion," 16; and
  distorted claims, 224
scientists: and advocacy, 29–31;
  consensus on climate change,
  128; and media, 24–26;
  motivations of, 13; public
  perception of, 10, 12; sense
  of wonder, 198–99;
  single-mindedness, 8–10
Seed (magazine), 26
senses, and health, 201–3
Skeptical Environmentalist,
  The (book), 229–30
skepticism, 28, 111, 128–30
smell, sense of, 201
smog, costs of, 92
social change, 219–22
social networks, 58
solutions, social nature of 271–72
Soviet Union, 27, 52
space, study of, 17–19
species: conservation of, and
  parks, 72–74; discoveries of,
  60, 75, 82; loss of, 38–41, 67
  (see also extinctions); number
  of, 14, 75; responses to
  disturbance, 78; study of, 14
Spence, Charles, 201–2
spirit bears, 75
spotted owls, 65, 69
sprawl, suburban, 153–55
Stern Review on the Economics of
  Climate Change, 100–102
subsidies, 91, 188
Survivor (TV show), 226–28
sustainability, 74, 42, 106
SUVS, 140–41, 146–48
Syngenta, 186

taxonomy, 80, 12–14
Taylor, Nick, 97
technology: and air pollution,
  196–98; effects on children,
  199–200; and environmental
  education, 215; human
  relationships with, 194–96;
  planned obsolescence
  of, 209–11
tobacco, and WHO, 222–23
Tory, John, 255
touch, lack of, 201–2
toxins: bioaccumulation of, 37,
  46, 70; in fish, 173; in food
  supply, 172–75; transport of, 46
transgenic organisms. See GMOs

uncertainties, 30
unconscious mind, and complex
  decisions, 259–61
Unimog, 142, 143–45
United States: bird species, 65;
  and climate change, 37, 42, 126,
  253; criticism of, 250–52; and
  GMOs, 185; and Kyoto Protocol,
  126; sustainability in, 42

video games, 202–3
"videophilia," 208
vultures, 63–64

waste, 36–38
watersheds, 88
weather, extreme, 93,
  119–21, 126–29
Wilson, E.O., 59, 81, 199, 212–13
wolves, 71, 76
wonder, sense of, 198–201
World Health Organization
  (WHO), 222–23
World Without Us, The
  (book), 106–7

ABOUT THE AUTHORS

DR. DAVID SUZUKI, cofounder of the David Suzuki Foundation, is an award-winning scientist, environmentalist, and broadcaster. David has received consistently high acclaim for his more than forty years of work in broadcasting, explaining the complexities of science in a compelling, easily understood way. He is well known to millions as the host of the Canadian Broadcasting Corporation's popular science television series, *The Nature of Things*.

An internationally respected geneticist, David was a full professor at the University of British Columbia in Vancouver from 1969 until his retirement in 2001. He is now professor emeritus with UBC's Institute for Resources, the Environment and Sustainability.

The author of forty-three books, David is recognized as a world leader in sustainable ecology. He lives with his wife, Dr. Tara Cullis, in Vancouver, B.C.

. . .

DAVE ROBERT TAYLOR is a writer and journalist and the president of Tipping Point Communications, a creative consulting agency. Formerly director of communications with the David Suzuki Foundation, Dave has received numerous corporate, provincial, and national awards for his writing; his work has appeared in the *Globe and Mail,* the *Vancouver Sun,* and other newspapers and magazines across Canada.

Dave has also written for television and film and is a past winner of the CBC's Signature Shorts screenwriting competition. Dave graduated with honors from the University of Victoria, where he received the Harvey Southam Award for his thesis on journalism ethics. Dave lives with his family in Tsawwassen, B.C.

THE DAVID SUZUKI FOUNDATION

The David Suzuki Foundation works through science and education to protect the diversity of nature and our quality of life, now and for the future.

With a goal of achieving sustainability within a generation, the Foundation collaborates with scientists, business and industry, academia, government, and non-governmental organizations. We seek the best research to provide innovative solutions that will help build a clean, competitive economy that does not threaten the natural services that support all life.

The Foundation is a federally registered independent charity, which is supported with the help of over fifty thousand individual donors across Canada and around the world.

We invite you to become a member. For more information on how you can support our work, please contact us:

THE DAVID SUZUKI FOUNDATION
219–2211 West 4th Avenue
Vancouver BC Canada V6K 4S2

www.davidsuzuki.org
contact@davidsuzuki.org

Tel: 604-732-4228
Fax: 604-732-0752

Checks can be made payable to The David Suzuki Foundation. All donations are tax deductible.

Canadian charitable registration: (BN) 12775 6716 RR0001
U.S. charitable registration: #94-3204049